OWNERSHIP OF THE HUMAN BODY

Philosophy and Medicine

VOLUME 59

Editors

EUROPEAN STUDIES IN PHILOSOPHY OF MEDICINE 3

Editors

The titles published in this series are listed at the end of this volume.

OWNERSHIP OF THE HUMAN BODY

Philosophical Considerations on the Use of the Human Body and its Parts in Healthcare

Edited by

HENK A.M.J. TEN HAVE

Catholic University of Nijmegen
Center for Ethics and
Department of Ethics, Philosophy and History of Medicine,
Nijmegen, The Netherlands

and

JOS V.M. WELIE

Creighton University,
Center for Health Policy and Ethics,
Omaha, Nebraska, U.S.A.

with the collaboration of

STUART F. SPICKER

Emeritus Professor of Community Medicine and Health Care,
School of Medicine, University of Connecticut Health Center,
Farmington, Connecticut, U.S.A.
and
Professor of Philosophy and Healthcare Ethics,
Massachusetts College of Pharmacy and Allied Health Sciences,
Boston, Massachusetts, U.S.A.

SPRINGER-SCIENCE+BUSINESS MEDIA, B.V.

A C.I.P Catalogue record for this book is available from the Library of Congress.

ISBN 978-90-481-5059-5 ISBN 978-94-015-9129-4 (eBook)
DOI 10.1007/978-94-015-9129-4

Printed on acid-free paper

TABLE OF CONTENTS

PART IV / OWNERSHIP OF THE BODY: THEORETICAL PERSPECTIVES

HENK A.M.J. TEN HAVE
JOS V.M. WELIE

MEDICINE, OWNERSHIP, AND THE HUMAN BODY

I. INTRODUCTION

One of the earliest illustrations of an autopsy performed on a human body is depicted in the image of a fourteenth-century physician, Guido de Vigevano, gently embracing the vertically positioned body he has opened with his lancet. The physician's facial expression reflects a sense of hesitation, even apology, for invading his fellow human being's bodily integrity [2].

Six centuries later we no longer consider it an affront to perform autopsies, or even completely to dissect human corpses. Indeed, cremation has become an acceptable funeral ceremony among most religious communities. We not infrequently extirpate organs from cadavers and living donors and transplant them into the bodies of desperate recipients. Immediately after death is declared we permit medical students to practice intubation techniques on the newly-dead and use human corpses to test the crash safety of automobiles. We clone human tissues in the process of producing pharmaceuticals, and some have considered cloning the human fetus to produce spare organs that are always in short supply.

The development of biomedicine is now associated by the public with changing views of the human body: it is no longer viewed as integral to the human individual (the Latin *individuum* meaning "what cannot be divided any further"). We have learned with the evolution of modern medicine to consider the human body, both as a corpse and as a living organism, as a machine-like construct, a useful instrument, even a marketable commodity. We purchase blood from "donors," trade organs in a world-wide market, patent human tissues and cell lines. The body (or at least its parts) has become *property*, owned by both the "incorporated person," and "incorporated firms."

H.A.M.J. ten Have and J.V.M. Welie (eds.), Ownership of the Human Body, 1–15.

II. OWNERSHIP OF THE BODY

The increasing influence of the moral principle of *respect for individual autonomy* in health care ethics has become associated with the popular image of the body as property [4]. It is assumed that an individual is an autonomous subject, that his or her body is his or her private property and that the person is thus the sovereign authority and possesses property rights over his or her body. Since an autonomous individual is said to "own" his body, he apparently possesses an exclusive right over it; he alone has it at his disposal. One can bequeath one's body to an anatomical institute, donate one's body parts for transplantation, or sell one's bodily materials in the open market. Indeed, bodies and body parts can be acquired and manipulated by others, though explicit permission may be required from their owners.

The use of property language has recently appeared in health care ethics to designate the locus of decision-making authority: the individual, as owner, retains control over his or her own body. In view of the ever increasing medical possibilities to intervene in the human body as well as the creation of new possibilities for using body parts in research and commercial enterprises, it has, according to some, become necessary to protect individuals from harmful and unwarranted paternalistic bodily interventions. At the same time, the concept of body ownership is morally as well as epistemologically problematic. If one distinguishes not only in thought between 'a person' and 'his body' this conflicts with the existential or lived identity of our bodies as our selves – embodied selves.

The concept of 'ownership', moveover, has recently been introduced in ethical debates with regard to medical genetics [5][20]. A major issue is the justification for the patentability of human genes. Making the human genome subject to property laws presupposes the relevancy and applicability of the ownership concept in this context. One basic question is: Is it morally permissible for anyone to "own" the human genome? If the answer is "yes" and ownership is judged appropriate in regard to genes, further questions arise that bear on who, precisely, is the owner of this genetic material. Is it the individual with a particular phenotype? Do geneticists, scientists, or corporate executives have a moral right to assert property claims over particular genes or nucleotide sequences? Or, is the human genome the "property of humankind" – "the common heritage of humanity," as is proclaimed in a recent UNESCO declaration? [23]

III. OWNERSHIP OF BODY PARTS

In the extant literature, the starting point for the analysis of the relation that obtains between the human body and the concept of ownership are the property rights that are characteristic for ownership. If a person is the owner of an object, this does not imply that he has one right only with regard to the object. Ownership is a complex collection of claim rights, duties, powers, and immunities. As a paradigm of ownership, reference is often made to A.M. Honoré's concept of 'full individual ownership' [14]. Honoré compiled a list of standard "incidents of ownership." Although the incidents are not, taken individually, necessary conditions for private ownership, they may, however, together be sufficient for full individual ownership. The standard incidents include: 1. the right to possess a thing; 2. the right to the exclusive use of a thing; 3. the right to manage it; 4. the right to its income; 5. the right to the capital, i.e., the right to alienate (transfer) the object, and the liberty to consume, waste or destroy the object; 6. immunity from expropriation; 7. the power to bequeath it; 8. the absence of term; 9. the prohibition of harmful use; 10. liability to execution; and 11. residuary character. This list is useful as a frame of reference in order to determine different modes of ownership.

With regard to ownership of the human body, it is possible to distinguish among three positions: (1) no ownership of the human body and its parts, (2) no full ownership of the body, but limited property rights with regard to body parts, and (3) full ownership of the body and its parts.

(1) No ownership of the human body and its parts

One way to acquire insight in the concept of the human body is to examine how we talk about our own body [17]. Sometimes we identify our self with our body ("He kicked me"), at other times we distance our self from our body ("He kicked my leg"). The use of the possessive pronoun suggests some mode of possession; but does "possession" amount to ownership? Is my body my property, or is it me – or both? The question really is whether some of the characteristics of ownership may be coherently applied to one's body. For example, can I disown or 'alienate' my body as I can 'alienate' my book? How did I acquire 'ownership' of my body? I did not labor for it, and I did nothing to deserve it. Is it a gift? How should I behave toward a gift? Can I dispose of it as I please? Considering the interpreting features of our body, we understand that we are of necessity embodied beings. Most hold that part of our dignity consists in acknowledging our fully-embodied existence. We are, moreover, not completely autonomous beings; we are part of a succession of generations. Many came before us, thanks to whom we exist; many will come after us. Our embodiment, therefore, is frequently considered a gift to be cherished and respected. According to L. Kass, some of the practices of modern medicine (e.g., reproductive technology, cosmetic surgery, organ transplantation) do not acknowledge the nature and meaning of bodily life, but have as their starting point the autonomy of the human person [17]. If the body is a gift that should be treated with respect, and if we are not completely autonomous beings, then selling and buying of organs, for example, would not be to recognize our bodily condition [24].

(2) No full ownership of the body, but limited property rights with regard to body parts

Authors defending this view argue that a person is not the full owner of his entire body, and that he, therefore, cannot claim a right to sell and destroy his body. The body is regarded as an integral part of the person. The authors, however, do not exclude the possibility of limited property rights with regard to body parts, because there is a certain similarity between body parts and ownable objects.

R. Harré introduced the concept of 'metaphysical ownership' to denote the internal relationship between a person and his body ([11], pp. 11-37;116-141). A condition for being this person is that I am embodied in this body. Because of my persistent body I am aware of my individuality/identity. This is an important characteristic that separates human bodies from things. I am, because of *this* body; without *this* body I would not be. Losing some of my private property will not have the same effect.

Although a person is intimately related to his body (as is expressed by the concept of metaphysical ownership), is there room for legal ownership of the body, or its parts? Legal ownership, according to Harré, includes a right to dispose of one's body (property) the way one chooses (if no other restrictions are imposed). The right of transfer is one of the rights an owner has with regard to his property. This right can be restricted, for example, if it concerns on object of aesthetic value. The legal right to dismantle or destroy it could be denied. Harré argues that the owner's right to dispose of valuable objects and the rights of persons with regard to their bodies are sufficiently similar to acknowledge that one has legal ownership with regard to some body parts. Disposal of body material in so far as it does not threaten the integrity of the body should, on Harré's view, be legally permitted. For example, the removal of hair, blood, and one kidney will not threaten the integrity of the human body. Harré does not make clear, however, whether he would allow the sale of all body materials that fall under legal ownership, or only the sale of some body material (thus further restricting the legal rights of the owner).

In developing his theory of property, S.R. Munzer analyzes rights with regard to our bodies ([18], pp. 37-58). His analysis, too, compares the rights we have with regard to things with those we have with regard to our bodies. Munzer defines 'property' as relations between people with respect to things. Property is thus 'a bundle of rights'. Do persons own their bodies? Assuming that the legal rights acknowledged by U.S. law are justified, then people do not own their bodies in the sense of full-individual ownership, but they may have some limited property rights over their body parts. To specify the rights people have over their bodies, Munzer analyzes the elements of the 'bundle of body rights'. He distinguishes between personal rights and 'strong' and 'weak' property rights. Not all rights a person has with regard to himself are property rights, some rights fall under privacy. The criterion employed to distinguish between personal rights and property rights is *transferability*. Personal rights are rights that protect interests or choices of a person, other than the choice to transfer. For example, the right of free speech is a right that a person has and can waive, but he cannot transfer it to someone else. Property rights are body rights that protect the choice to transfer. If a person is only permitted to donate some body material, then he has a 'weak' property right with regard to this body material. 'Strong' property rights permit and involve the transfer of body material in open market exchange.

According to C.S. Campbell, three elements of the 'property paradigm' can be found in the literature on biomedical ethics [4]. First, a right of territorial integrity: people have a right of bodily integrity, because they *are* and *possess* their bodies. Physicians are generally not permitted to operate on patients without their fully informed consent. Second, developments in modern medicine have made it possible to separate a whole range of 'parts' from the body. In addition to hair, blood, sperm, certain tissues, and germ cells, organs are transferred from patient to patient. In terms of property: organs can be alienated (transferred) and acquired. This raises the question of control and legal possession, the third element of the property paradigm. If the body is property, who is its owner? Who has the legal right to possess (or exclusively control) the body? To assert that a person is the owner of his body and its parts is to acknowledge his right to control what happens to them.

However, the observation that elements of the property paradigm can be applied to rights people have with regard to their bodies (right to bodily integrity) or to procedures taking place within medicine (organ transplants), provides no justification for the restricted use of the property paradigm with regard to the human body. Analyzing what is essentially a religious concept (stewardship) and a secular conception (self-ownership) of ownership of the body may help further to clarify this issue. In both of these conceptions alienation (transfer) of body parts is possible. Within some Western religious traditions even the sale of some body materials would not be morally objectionable. Campbell, however, has some reservations with regard to the commercialization of body parts. On his view, safeguards must be established to prevent treating the human body as mere property, for this would be to compromise (deny?) our embodied existence.

A somewhat different approach is presented by T.H. Murray [19]. Recognizing the value of the body, both for oneself and for others (e.g., organs for transplantation; tissues for research), there are at least two ways to view body parts: they may be seen as property that can be bought and sold, or they can be construed as gifts. Murray argues that at least some body parts should not be regarded as private property, but as gifts. According to him, gifts are important in establishing and maintaining moral relationships among people, because gifts create obligations for donors and recipients. Gifts to strangers (e.g., blood or organ donations) are important to maintain authentic relationships within society and for respecting specific human values. They underscore our interdependence and the value of solidarity and human dignity.

This does not mean, however, that all body parts must be treated as gifts, as items that should not be bought or sold. After all, not all body parts are equally essential for human life. Murray mentions urine, nails, and hair. The sale of these body materials would hardly threaten human dignity. Other body parts, however, should be regarded as gifts, primarily due to their importance in sustaining the lives of members of society. For Murray, some body material is fully owned (i.e., one has the right to sell the material), but other body materials are only owned in a limited sense (i.e., one has no right to sell them, only to donate them).

(3) Full ownership of the body and its parts

Some authors recognize legal ownership of the human body, although restrictions on what a person will be permitted to do with his body are still possible. L.B. Andrews maintains that people's body parts are their personal property [1]. The reason for this acknowledgement is twofold: (1) the fact that in U.S. law the body is sometimes treated as property; (2) the 'property approach' is the best way to protect the interests of people concerning their own body parts. It will enable them to control what happens to their body and its parts. According to Andrews, there is no reason to prohibit the sale of all body parts. A prohibition is justified for the donation or sale of non-generative body parts, where donation entails the death of the donor. For Andrews, allowing the sale of body parts will not necessarily lead to the public's judgment that human beings are merely commodities. A safeguard against this danger is that only persons themselves are allowed to treat their body parts as property; others do not have the right to treat me as property.

Another advocate of full ownership of one's human body is H.T. Engelhardt ([7], pp. 127-134; 365-6). For him, the fair allocation of scarce healthcare resources requires us to know what would be a morally justified allocation; to know this it is necessary to know who owns what and in what way. Engelhardt's theory of ownership incorporates the views of Locke and others on the acquisition of ownership. For Locke, a person acquires ownership of a thing (land) by mixing his labor with it – one takes possession of a thing by grasping it, forming it, and by marking it as one's own. The classic example of possession is one's 'possession of oneself' (we form and use ourselves). How do we acquire ownership of things other than ourselves? Partial ownership of other persons we acquire by consent, or because we are the producers of them (our children). Ownership of things we acquire by grasping, forming, marking, and laboring on them. By these actions we extend our person in the thing and thereby bring it within the sphere of mutual respect (i.e., others may not interfere with the thing without our consent).

Apart from private ownership, there are two other forms of ownership, namely communal ownership and general ownership. Communal ownership are the resources that are brought together through a free, common endeavor of the members of a community. General ownership is the right of every person to the rough material of the earth. Only the communal resources can be allocated to health care projects by common consent. But persons with private property will always have the right to use it in the way they see fit. Physicians, for example, have the right to sell their services outside a national health care system, because they are the owners of their talents.

In this theory of ownership a person is the private owner of himself and therefore of his body, its parts, and his talents. A free individual has the right to dispose of his property (and thus himself) as he pleases. The state has no right to interfere with the transactions of free individuals. A prohibition on the sale of one's organs, therefore, would not be morally justified. It seems that Engelhardt would allow the sale of organs following the removal of which would have as a consequence the death of the donor. Generally, however, most persons would prefer to prohibit the sale of vital organs, like the heart. But the philosopher J. Harris does not see why he should not be permitted to give, or even sell, his heart (and therefore his life) if that is what he wills to do, and he fully knows what he is doing [12].

IV. THE CONCEPT OF OWNERSHIP OF THE BODY

These various views regarding ownership of the human body and body materials have as their principal focus the possible sale of body materials, such as organs for transplantation. In the context of genetics, the issue of ownership is not so much related to the matter of selling genetic material and information, as it is the moral right to prohibit others from having access to one's genetic information/material.

The survey of different positions with regard to ownership of the human body shows that the concept of 'property' is easily applied to the human body, although none of these conceptions really amounts to full, individual ownership. Nevertheless, it is clear that there is a range of possibilities with regard to ownership of the body. At one extreme there is opposition to owning the body and its parts; at the other, there are those who argue that the selling of one's organs (even vital organs) is morally licit. Between these two extremes some argue for limited property rights with regard to some body parts.

In this volume, various dimensions of the problem of ownership of the human body are examined. Emphasis is on the philosophical and moral (and at times legal) dimensions. Part I contains various cases and examples from medical practice where the issue of ownership of the body arises. Bela Blasszauer, a Hungarian bioethicist, focuses on the use of the human body after death [3]. One of the oldest medical practices is the dissection of the body in order to study its structures and functions. The development of anatomy as a medical discipline was in fact one of the first signs of the scientific turn of modern medicine. But this use of the body – to obtain medical knowledge – also involves a tension. On the one hand, the human body can be used as a source of useful information for other human beings (e.g., for the purpose of education, improvement of diagnoses, increasing experimental knowledge of potential value to other patients with similar diseases and conditions, to enhance quality control of therapeutical interventions, etc.) The body, therefore, is a source of education and an instrument for enhancing the progress of medical science. Body materials can also be used to develop products that in the end save the lives of other human beings. Indeed, the development of knowledge is a moral value contributing to the relief of human suffering; to this end the human body ought to be regarded as a valuable source for obtaining such knowledge. On the other hand, there is a general consensus that the human body ought not to be treated as a mere instrument, used simply to obtain knowledge for its own sake. Due to the special relation between body and person, the intrinsic value of the human body is ever-present. In nations like Hungary, the human body is considered property of the state; the medical profession is free to perform autopsies without being required to obtain an informed consent from the families of deceased persons. If the person himself has not objected in advance to autopsy, his or her consent is presumed and, as Blasszauer states, "practically anything can be done to the body." This "freedom" can lead to rather dangerous behavior toward bodies. Indeed, Blasszauer offers several examples. To change this situation is quite difficult, since it would require a revised concept of ownership, where one argues that persons themselves are the owners of their bodies, and therefore only they have the right to decide what happens to their bodies after death.

The remaining contributions to Part I focus on the use of body parts. Friedrich Heubel addresses the question of how to treat parts of the body [13]. In Germany, in 1993, the Federal Court awarded damages to a plaintiff whose sperm had been "negligently discarded" by a sperm bank.

This ruling implied that sperm are part of the body that had been separated from the body, but remained part of the plaintiff's body. In its formal opinion, the Court tried to specify the legal status of the plaintiff's body parts. Parts of the body once separated from the body are usually considered things; the person from whom these parts have been obtained retains, however, property rights over these former body parts. However, this general rule was modified in this specific case. The German Court argued that parts of the body which have been separated from the body but are destined to be reunited with the same body are NOT things, because they continue to belong to the body (e.g., human ova fertilized *in vitro* and destined to be implanted). On the other hand, when the same separated parts are destined to be united with another person's body, they are things (e.g., donated blood). The Court ruled that human spermatozoa are an exceptional case: they have been separated from the body in order to achieve procreation in a woman's body, but they still belong to the body of the sperm donor, because they are destined to fulfill a particular function of his body. Sperm are a functional unit of the body from which they originate; in fact they embody the possibility to perform a typical bodily function of the plaintiff. Underlying this reasoning is a redefinition of the body as the totality of those organs and functions which a person can use as a means of achieving self-chosen ends. The human body in this perspective is a functional organism that is used by the person to realize particular values and life ends.

One extremely important body material in present-day medicine is blood. Donation of blood is also one of the oldest ways of transmitting body parts from one person to another. From a moral point of view, as Hub Zwart argues, blood donation is traditionally regarded as an act of altruism [27]. But different systems have developed: commercial and donor systems. Debates concerning whether payment for blood is desirable turn on two kinds of ethical considerations: (a) the moral consequences, i.e., it may lead to inequalities between the poor and the rich, for it may encourage transactions of other body parts, promoting further commodification and commercialization of the body and its various parts; commercial systems seem also to involve reduced safety; and (b) the moral value of donation itself. Gifts affirm the relationship of solidarity we have with our fellow human beings in society; they preserve a communal sense. In discussing these approaches, Zwart distinguishes between two moral "vocabularies" in the moral debate concerning blood donation: a gift vocabulary and a commodity vocabulary. He shows that they entail different views on personhood and ownership. The basic question, then, is whether we prefer to regard the human body as property, as a thing at our disposal, or as something entrusted to our care.

Similar questions arise when the debate focuses on the use of body parts, as Wim Dekkers and Henk ten Have suggest, in the context of medical research with human body parts [6]. One problem is what we shall take to be a 'body part'. The authors argue that two characteristics are essential: integration of the part within the entire body (organism), and origination of the part in the human body. Collection, use, and storage of human body parts are necessary for modern medicine to carry out its range of aims and tasks. Various examples and cases are discussed, particularly in the context of biomedical research with human body parts. These cases highlight numerous ethical concerns, for example, issues of privacy and informed consent. But a more fundamental issue underlying these concerns is the question of the moral status of human body parts: how far does one's sovereignty extend with respect to his or her body parts?

Part II is analytical: the authors first examine the history of thoughts on ownership of the body, and then present a conceptual analysis. Diego Gracia raises the following question: who is the owner of the human body [10]? As the history of slavery shows, it was commonly thought that at least some people should not have a property right over their own bodies. In Western philosophy, three different positions can be distinguished concerning ownership of the human body: (1) God is the owner of the human body. This view prevailed in Greek and early Christian philosophy. Individual persons are not the owners of their bodies, but the administrators or stewards, who only have the right to dispose of the body for the *good* of others. (2) The individual person owns his body. This is the view of liberal philosophers (e.g., John Locke). The body is the first property of the human being. Only he can decide what will happen to it. Kantian philosophy developed a slightly different view: the human being is indeed the owner of his body, but he is not entitled to dispose of it as if it were *merely* his property; he would on that account be treating himself as a mere means or object. (3) The principal owner of the body is society; this is the view of socialist philosophy. The human body is not only private property, but it is also society's property since the body always has two dimensions: it is (1) consumer and producer, thus being private and individual, and at the same time it is (2) public and social. It is argued from this perspective that the human body is first of all a public and social body, and then it becomes personalized as an individual body. After analyzing these three views, Gracia then proffers an integrative model for understanding issues surrounding the ownership of the human body. He argues that the body always has two dimensions: the public and the private; both need to be integrated. He defends the theory that health is equal to possession or appropriation of the body. Again, we have two levels: lack of dispossession (the public level), and perfect possession (the private level).

The human being is also two-dimensional: it is personal and individual; my body coincides with my personal identity. At the same time, my body belongs to the human species and only survives in a social context.

Zbigniew Szawarski further analyzes this conceptual issue [22]. He explicates the difference between parts of human bodies and mere things. He takes, as an analogy, the situation of a blind man who uses a white stick to orient himself in space; comparing it to the eye that helps others to orient themselves. Things and instruments can be sold and bought; there is therefore always a boundary between an instrument and the living body. Body parts such as the eye cannot be used as instruments; they are not things. Therefore they are not property. Szawarski rejects Locke's view that I am the owner of my body. My body parts belong to me or *appertain* to me in a non-possessive way. The eye is subjectivized – endowed with the human subject. It can be property only after I die. Continuing this line of argument, the human body becomes a thing only after death; only then may it become someone's property. Between the body and the person there are a variety of relationships, e.g., sometimes my body is a hinderance; at others times my body is "absent," i.e., it seems that I no longer live in it. In elaborating upon these relations, Szawarski also uses the two vocabularies discussed by Zwart: gift is the language of solidarity; selling is the language of commodity.

Part III focuses on statutory considerations. The authors compare ways in which societies and legal systems deal with ownership issues. Jos Welie and Henk ten Have discuss whether the human body can, under Dutch law, be the object of a legal relationship [25]. They conclude that this is not the case; there is no indication that in Dutch law the human body is regarded to have inherent value independent of the person whose body it is. The situation, however, is different for body parts; when these are separated from the body, they may become objects of legal relations, including ownership. Once separated, they are like things owned by the person from whose body they originated. However, although it is agreed that the person who incorporates or has incorporated the body has power or control over his or her body parts, the moral principle of respect for personal autonomy rather than the concept of ownership is employed (under Dutch law) to justify this control.

The legal situation in France, described in detail by Anne Fagot-Largeault, allows citizens to donate body parts as an anonymous gift [9]. Indeed, the use of human subjects in medical research has been extensively debated in France. An important determinant in the moral debate was the conviction that the human body ought not be regarded as a thing, and therefore ought not be viewed as mere property. The human body and its parts can thus not be marketed or sold, because under French law, I *am* my body. The basic concern here, apparently, is 'human dignity': human beings may not sell parts of themselves because it is personally degrading. In the French legal tradition, therefore, persons are not the owners of their bodies.

In Part IV, different theoretical perspectives on the human body are presented and analyzed. Fr. Kevin Wildes argues that it is no longer possible to endow special meaning to the human body, given our postmodern predicament [26]. From a libertarian position, which he takes to be the only defensible moral position, bodies and body parts may be used only with the explicit *permission* of the person whose body it is. The secular state, if it wants to avoid the particularity of moral visions, as it should, should not be permitted to interfere with the transactions of free individuals as private owners of their bodies; there is no reason, furthermore, to prohibit the commercialization of body parts under this construal. In fact, Wildes explicitly acknowledges and further defends Engelhardt's position – one that recognizes full ownership of the body.

A quite different perspective is advanced by Paul Schotsmans [21]. He argues that medical morality should assume the fundamentally personal character of the human individual, which leads to a broader and richer view than one's right to self-determination. Personalist ethics emphasizes not only the "unicity" of each person, but also the significance of communication and human solidarity. This moral theory presupposes a specific anthropological view on the human person and human body: the body is part of the integrated subject that is the person, while at the same time it is also part of the material world. Although Schotsmans does not elaborate, his perspective apparently leads to the view that one does not own one's body and its parts. This view is based on two ideas crucial to personalist ethics, *viz.* that a person is his body, and that a person is not completely autonomous.

The same theoretical position, although from a different point of view, is defended by Uffe Jensen [16]. He analyzes the growing tendency in Danish society to consider the body and its parts as the property of the individual. He argues that this tendency should not be explained on the basis of personal autonomy, but rather as a rhetorical device to articulate group interests. In opposition to the libertarian perspective, influenced by Locke, Jensen adopts an Hegelian approach, and argues that the concept of property is philosophically inappropriate when discussing the human body and body parts.

Analyzing the case of organ donation, Franz Joseph Illhardt develops a deontological perspective on the human body, arguing that although the human body belongs to the person himself, it should not belong to other persons, since that would obfuscate if not violate the principle of respect for autonomy which recognizes the fundamental agency of each individual [15].

In the closing essay, Martyn Evans addresses the utilitarian moral perspective [8]. He argues that utilitarian theory has serious, though often unnoticed, limitations; it cannot, theoretically, assist in the resolution of the problem of who owns the human body; but it can, practically, argue for the uses of human bodies and body parts. Utilitarianism will typically focus on justifications for the view that bodies can be owned because of the benefits that using bodies will bring to someone, or to others.

It is apropos, given the contributors' rather serious treatment of the theme of this volume, to mention what one might consider THE extravagant "argument" for body ownership: On display at University College, London, the visitor can view the "remains" of the founder of utilitarianism, Jeremy Bentham, for they are exhibited in university hall; visitors surely wonder who or what they are seeing. Bentham surely considered himself the "owner" of his body; now, however, "he" is the property of the institution established by his philosophical disciples....

BIBLIOGRAPHY

1. Andrews, L.B.: 1986, 'My Body, My Property', *Hastings Center Report* **16**, 28-38.
2. van den Berg, J.H.: 1959, *Het menselijk lichaam. Vol. I: Het geopende lichaam.* Callenbach, Nijkerk, the Netherlands.
3. Blasszauer, B.: 1998, 'Autopsy', in this volume, pp. 19–26.

4. Campbell, C.S.: 1992, 'Body, Self, and the Property Paradigm', *Hastings Center Report* **22**, 34-42.
5. Danish Council of Ethics: 1994, *Patenting human genes. A report.* Copenhagen, Denmark.
6. Dekkers, W. and ten Have, H.A.M.J.: 1998, 'Biomedical Research and Human Body Parts', in this volume, pp. 49–63.
7. Engelhardt, H.T.: 1996, *The Foundations of Bioethics*, 2nd ed. Oxford University Press, New York, NY/London, UK.
8. Evans, M.: 1998, 'The Utility of the Body', in this volume, pp. 207–226.
9. Fagot-Largeault, A.: 1998, 'Ownership of the Human Body: Judicial and Legislative Responses in France', in this volume, pp. 115–140.
10. Gracia, D.: 1998, 'Ownership of the Human Body: Some Historical Remarks', in this volume, pp. 67–79.
11. Harré, R.: 1991, *Physical Being*. Basil Blackwell, Oxford, UK.
12. Harris, J.: 1992, *Wonderwoman and Superman*. Oxford University Press, Oxford, UK.
13. Heubel, F.: 1998, 'Defining the Functional Body and Its Parts: A Review of German Law', in this volume, pp. 27–37.
14. Honoré, A.M.: 1961, 'Ownership' in Guest, A.G. (ed.): *Oxford Essays in Jurisprudence*. Clarendon Press, Oxford, UK, pp. 107-147.
15. Illhardt, F.J.: 1998, 'Ownership of the Human Body: Deontological Approaches', in this volume, pp. 187–206.
16. Jensen, U.J.: 1998, 'Property, Rights and the Body: The Danish Context – A Democratic Ethics or Recourse to Abstract Right?, in this volume, pp. 173–185.
17. Kass, L.R.: 1985, 'Thinking About the Body', *Hastings Center Report* **15**, 20-30.
18. Munzer, S.R.: 1990, *A Theory of Property*. Cambridge University Press, Cambridge, UK.
19. Murray, T.H.: 1987, 'Gifts of the Body and the Needs of Strangers', *Hastings Center Report* **17**, 30-38.
20. Pompidou, A.: 1995, 'Research on the Human Genome and Patentability – The Ethical Consequences', *Journal of Medical Ethics* **21**, 69-71.
21. Schotsmans, P.: 1998, 'Ownership of the Body: A Personalist Perspective', in this volume, pp. 159–172.
22. Szawarski, Z.: 1998, 'The Stick, The Eye, and Ownership of the Body', in this volume, pp. 81–96.
23. UNESCO: 1995, 'Revised Outline of a Declaration on the Protection of the Human Genome', *Eubios Journal of Asian and International Bioethics* **5**, 97-99.
24. U.S. Congress, Office of Technology Assessment: 1987, *New Developments in Biotechnology: Ownership of Human Tissues and Cells – Special Report.* U.S. Government Printing Office, Washington D.C. (OTA-BA-337).
25. Welie, J.V.M. and ten Have, H.A.M.J.: 1998, 'Ownership of the Human Body: The Dutch Context', in this volume, pp. 99–114.
26. Wildes, K.: 1998, 'Libertarianism and Ownership of the Human Body', in this volume, pp. 143–157.
27. Zwart, H.A.E.: 1998, 'Why Should Remunerated Blood Donation Be Unethical? – Ethical Reflections on Current Blood Donation Policies and Their Philosophical Origins', in this volume, pp. 39–48.

PART I

OWNERSHIP ISSUES IN CLINICAL CARE AND BIOMEDICAL RESEARCH

BELA BLASSZAUER

AUTOPSY

I. INTRODUCTION

Despite wars, terrorism, ethnic massacres, and international murders and killings, a discourse on respecting the human body even after death should not be understood as meaningless. We can surely hope the time will come when peace and tranquillity prevail throughout the turbulent places on our planet, when science in general and medicine in particular can concentrate on service to humanity, and scientists can work in an environment of security and hope. In the meantime, it will remain unimaginable in our secular world to demand reverence and respect for dead bodies since so many continue completely to disregard the very lives of their fellow human beings.

It is a straightforward empirical fact that human beings are born and die; they die without wars and catastrophes. The sentient, rational being is aware of the fact that his life is finite. One can imagine one's own death and the reaction of one's loved ones to this final event. Most of us would not like to be treated as an object even after our death because we believe such treatment would violate our essential dignity and humanness. Respect for the dead is often understood in terms of respect for the living being. A human cadaver is the symbol of the human person. Ancient norms prohibiting the desecration of the dead are practically universally demanded based on long-standing tradition and cultural beliefs. Although cadavers still possess human significance, an autopsy can often serve the common good by, e.g., contributing to the development of medicine, training future physicians, and serving public health and safety, following determination of the cause of death, which provides subsequent control over therapeutic measures. Regardless of what a dead body is, its desecration and/or mutilation constitutes a moral offence. While Western nations typically legislate stringent consent requirements governing the use of the dead body, in post-Communist countries and countries with developing democracies or no democracy at all, the dead body is generally

H.A.M.J. ten Have and J.V.M. Welie (eds.), Ownership of the Human Body, 19–26.

considered the property of the state, with the medical establishment acting as trustee.

II. CADAVERS AS MERE MEANS

In Hungary, for instance, where democracy is still in its cradle and respect for human rights has only recently become a goal of the government and public policy, a decree of 1984 issued by the Minister of Health, and modified in 1987, permitted the virtually unlimited use of human bodies: routine autopsy, scientific research, organ transplantation, and drug production. In former communist countries such as Hungary consent is not required to perform an autopsy on a dead body. Practically anything can be done to the body of a deceased person who did not file a formal protest against such activity while alive. The policy of presumed consent places no straightforward limit on the various uses of human cadavers. However, very few Hungarians are aware of this legal trap, since the inaction of this law neither by appropriate public information nor by precedent established ways and means for citizens to express an opinion. In the United States the situation has been quite different. As Arthur L. Caplan writes: "Americans have been [and the report of the Task Force created by the National Organ Transplantation confirms] and remain staunchly committed to free, voluntary choice and altruism as the appropriate moral foundations for cadaver donation" ([2], p. 79). The majority of the Hungarians would surely agree with the American practice, and if given the choice they, too, would prefer free, voluntary choice as the proper moral basis for use of their dead bodies. Hungarians, as persons, have dignity and moral worth that ought to be respected as such serious moral concerns conflict with the possible benefit derived from autopsy, transplantation, and research with human subjects.

The apologists for the Hungarian legal system argue that the Health Act is one of the best in Europe, quite modern, and that post-mortem examinations are one of the great achievements of human culture. This legal 'solution' is also very much favored by transplant surgeons and pathologists. It is argued that the greatest moral deed is to help to save the life of a fellow human being even after one dies by having one's organs, tissues, and bones accessible for research and transplantation. One constantly hears the Latin proverb 'Hic mortui vivos docent', meaning, 'here the dead are teaching those who are alive'. No one seems to be troubled by the numerous articles about the poor hygienic state of pathological institutes; finding human body parts scattered around in cemeteries; or the national scandal that shocked Hungarian citizens a few years ago. It concerned pituitary glands that had been surgically removed from 80 to 100 thousand human cadavers over some years, and sold to foreign firms or, as has been alleged, exchanged for supplies of human growth hormone. The Hungarian Health Act of 1972, however, forbids any kind of reimbursement for organs or tissues removed from dead bodies. Nevertheless, the national outrage was not enough to change the situation. Contradictions remain in the laws which allow some state-owned companies to profit by trading and selling human parts, despite the Health Act and the consensus among lawyers that the dead human body is not commercial property.

The author, whose job is to follow, monitor, and instigate changes in the moral and legal aspects of medicine and health care, has not yet heard about one person filing a protest against autopsy or organ/tissue removal after his death. Although theoretically everyone has the right to do so, practically speaking, the opportunity to protest effectively is unavailable. In a country where 'presumed consent' regulates the fate of one's body after death, one is forced to wonder whether the absence of protest was the genuine choice of the deceased, or whether there was no other alternative. Indeed, how can one have his wishes known?

A survey conducted among medical students and health professionals in 1992 concludes that out of 403 respondents only 68 ever heard of *presumed consent*. More than half stated that they had not heard of it even after the concept was explained to them in detail. Those who knew about the doctrine and its actual application were typically health professionals. The survey also demonstrated that more than half of the respondents would be willing to donate their organs to save the life of another human being. There were, however, many who would not. The reason mostly given was not their unwillingness to help others, but their profound mistrust of the medical and legal establishment [1].

The present public policy in Hungary concerning the disposition of the body of the deceased virtually gives a free hand to physicians to do with the dead body whatever they judge to be instrumental for saving and improving lives. This means that there is only one overarching goal, a single goal that is achieved through autopsy, organ transplantation, drug production, and tissue and bone banking. Other possible goals, such as respect of the interests and values of the deceased or of his/her family, or the encouragement of altruism and solidarity, are generally ignored. Furthermore, one is forced to wonder how this narrow-minded policy will effect the public trust in the medical profession. Perhaps even those who are supportive of autopsy, organ transplantation, biomedical research, and of finding new pharmaceuticals will change their minds; so when the time comes for autonomous decisions about donating their bodies for legitimate purposes, they will simply refuse.

The autopsy can be a very useful technique for teaching, evaluating medical practice, and diagnosing the cause of death, as well as for advancing medical science. The usefulness of dissecting the human cadaver has been known for centuries; fortunately, the time has gone when cadavers were obtained by grave robbing or murder. Although it is difficult to accept the claim that it is necessary to carry out a vast number of post-mortem examinations, or that "autopsy is such a procedure with which nothing can compete in assessing the efficacy of the healing art and to promote its progress" (as a pathologist stated in a local journal ([3], p. 10), still, no one doubts that it is, indeed, important. Nevertheless, autopsies can be carried out with due respect for the dead, that may include the complete reconstruction of the body before burial so as to avoid mutilating the body. However, some counter arguments can be proffered against placing too much emphasis on autopsy, especially if it is undertaken without obtaining informed consent; or, if it is carried out against the expressed wish of the relatives who report that the deceased would have objected had he/she been given the opportunity to deny permission. Even given the fact that post-mortem examinations are completed in great numbers, in Hungary, it does not seem to be the case that medical progress has advanced significantly. In fact, Hungary's morbidity and mortality statistics reveal exactly the opposite. It seems that pathologists are also more and more occupied with tissues, histological materials removed from living persons for examination, and helping clinicians arrive at the correct diagnosis in order to administer therapeutic measures. There are also new diagnostic techniques that can replace autopsy findings; they are probably even less costly. The above-quoted pathologist states that "the patient's physician and the pathologist agree – against the wish of both patient and relatives – that autopsy is necessary; thus, in such cases it is quite obvious that the 'professional views' have priority since they are serving the interest of the society" ([3], p. 10). While this is often said in Hungary, it seems to contradict another often heard complaint pertaining to the lack of close collaboration between pathologists and clinicians. Regardless of the quality of collaboration between the two specialists, the notion not to retard progress in medicine, prevails. This clearly overrides several moral values that have prevailed throughout Hungary's social, cultural, and religious traditions. Not to mention the fact that the sacrifice of present people and their system of beliefs for the benefit of future generations should be based on convincing arguments. Thus the 'debate' on the ownership of the body has already been decided on the basis of the above utterances; namely, if physicians feel, for example, that an autopsy is necessary because of the 'interest of society', then no one can act contrary to this principle, since the laws make it possible for physicians to go ahead and serve the 'interest of society'. Thus any moral doubt raised concerning the procurement of organs, for example, is viewed with suspicion, since what else could be more important than to save the life of those who are in urgent need of organs, or who would benefit from the discovery of a wonder drug? After all, as has been voiced by many prominent Hungarian physicians, there is no ownership over the dead body; if there is no owner, then it is quite obvious that there is no one who has the right to refuse to use human body parts to save the lives of others. If such an 'irrefutable argument' exists in support of involuntary organ procurement, then what can one say about carrying out autopsy within this ethical framework? If the medical profession spreads the idea that the ends justify the means and if it misinforms the public about the status of cadavers in other western nations, then there is very little hope that our society's interest in respecting dead bodies (together with the traditional cultural values and sentiments of the survivors) will soon be realized.

III. AN ATTEMPT TO CHANGE THE PRACTICE OF AUTOPSY

An ethics committee at a Hungarian medical university has worked out unprecedented ethical guidelines for handling dead bodies. It has put forth general principles that very much differ from the present legal rules and from actual practice. Therefore, it must be considered something revolutionary. It takes into consideration the interest of the individual citizen as much as the interest of society. The guidelines place the particular human being at the center of attention, not simply human life in general. It states that each human being is the natural owner of his/her own body; that each person has the right to decide what happens with his or her body after death. In cases where there is no living will, those who take care of the person's burial have the right to make such a decision. The shipping, storing, and handling of cadavers should be significantly improved by these guidelines and all the necessary means for exercising respect toward the dead should be available. There ought as well, perhaps, to be strict control to ensure that those persons responsible for handling the deceased uphold these moral, professional, and hygienic rules. Continuous education of these people is also a necessary requirement. It is highly desirable to change the present practice in such a way that the physician must obtain the informed consent from the relatives for autopsy, and not the other way around – where the relatives had formally to appeal to physicians through a rather bureaucratic procedure *not* to carry out an autopsy. It is also quite timely that an acceptable balance can exist between serving the interest of both the individual and society. This can be achieved in many ways. For example, if physicians inform the public about the usefulness of autopsies and publicly express gratitude to those who consent to this procedure. It is also necessary to conduct a cost-benefit analysis and to utilize the results of the autopsy. The guideline also requests that the university creates acceptable conditions for receiving the relatives of the deceased in a culturally and morally acceptable milieu. The ethics committee noted that this area had been neglected in the past, and it was now time to pay more attention to it. Regardless of how 'revolutionary' this guideline is in our circumstances, thus far it has produced no nation-wide response; but it did result in considerable local (town and county-wide) improvement in the practice of conducting autopsies.

IV. AUTOPSY AND MEDICAL STUDENTS

As mentioned earlier, autopsy is important in teaching medicine to future physicians. Students can learn a lot about the medical profession, and also about themselves – about life and death in general. The undergraduates' first encounter with human bodies is usually a long-lasting emotional experience. Most of these students are not prepared adequately to face the naked truth of mortality and actively to participate in the dissection of a dead person. The reaction of medical students to dissection has been surveyed by J.C. Penny. According to her, 23% of the respondents suffered from either nausea, fainting, loss of appetite, sleeplessness, or nightmares during the first weeks of dissection; 38% reported changes in attitudes after the dissecting experience. The changes included an unwillingness to donate their own body for medical purposes, an unwillingness for close relatives to donate their bodies, and a new awareness of their own mortality [4].

Unfortunately, the lack of adequate preparation of students, and at times the permissive or indifferent attitude of their teachers, can and does lead quite often to childish behavior in the autopsy room, and to a violation of the ancient norm of reverence and respect. An introductory lecture about proper conduct before, during, and after the autopsy should precede the actual dissection. Students, perhaps, ought to consider the autopsy room to be a more sacred place. Schiedermayer states that "the way students treat cadavers reveals much not only about the kind of doctors they will become but also about their view of the [dead] human being" ([5], p. 3). Many laymen and even some health professionals believe that if the time provided to pathology is excessive, then the more students will become detached from their living patients. Again, the right balance should be sought. It is hard to imagine that those students who fail to respect the dead, who treat their bodies as disposable objects, and perhaps later on as the source of raw materials, will ever come to respect their patients.

V. THE MEMORY OF THE DECEASED

Witnessing the dying and the death of one's loved one can have a great impact on one's memory of the deceased. The last sight of the dead body is especially imprinted in our memories, perhaps deeper than anything else that we had earlier encountered. Thus it makes a great difference how we see our beloved friends or relatives prior to the funeral rites. Much depends on the reconstructive efforts of pathologists, following the autopsy. Any sign of possible mutilation of the body may induce in us aversive feelings that might remain throughout our lifetime. It may be called 'psychological damage' or just a fantasy about our own final fate; in either case can one deny that some kind of harm has been done. If the disfiguration of the body by the removal of organs, bones, and skin has been undertaken without consent, without the conscious display of respect or altruism, then great harm will have been done both to the individual and society; moral stability will have been undermined.

Coercion, manipulation, misguidance, and deceit always weaken social cohesion even though this may not be obvious at times; in the long run the consequences of such behavior will be clearly visible. In despotic societies the question of ownership of the body is hardly raised. The body – as everything else – belongs to those who possess political power. Where living human beings are not respected and human life has hardly any value, what happens to human bodies after death becomes or remains a trivial matter. Despotism has, fortunately, virtually disappeared, and moral, political pluralism is emerging. Thus there is hope that both the living and the dead will be respected, receive equal treatment as human beings whether dead or alive, and realize the possibility of genuine altruism and expanding solidarity.

Medical University of Pecs
Pecs, Hungary

BIBLIOGRAPHY

1. Blasszauer, B.: 1992, 'What does presumed consent mean?', *Unpublished survey.*
2. Caplan, A.L.: 1991, 'Assume nothing: The current state of cadaver and tissue donation in the United States', *Journal of Transplant Coordination* **1**, 78-83.
3. Kadas, I.: 1991, 'When the dead are teaching the living', *Uj Dunantuli Naplo* August 10, 10.
4. Penny, J.C.: 1985, 'Reaction of Medical Students to Dissection', *Journal of Medical Education* **60**, 58-60.
5. Schiedermayer, D.L.: 1992, 'Intimate Revelations: An Introduction to Cadaver Ethics', *Journal of Clinical Ethics* **7**(2), 3-4.

FRIEDRICH HEUBEL

DEFINING THE FUNCTIONAL BODY AND ITS PARTS: A REVIEW OF GERMAN LAW

Do sperm which have been separated from the body still belong to the body? This question had to be decided by the German Federal Court in November, 1993. The court answered this question in the affirmative, and awarded damages in the amount of 25000 DM to a plaintiff whose sperm had been discarded by a sperm bank.[1] According to German law, this signalled that discarding sperm against the wishes of the donor can be regarded as bodily injury. In other words, the Court considered sperm a part of the body, which had been separated from the body, as a part of that body, though not as an arm would be considered. Thus, implicitly, the Court defined what a human body is in law. For this reason the decision is worth discussing from the perspective of medical ethics.

In the following four sections I report the case history of the plaintiff; give some explanations of the German legal position; describe the argumentation of the Court; and, finally, reconstruct the Court's notion of the human body in terms of moral philosophy (by referring to Kant).[2]

I. THE CASE HISTORY

At age 31, Mr. B. was diagnosed as having carcinoma of the urinary bladder. He was advised to have his entire bladder, including the prostate, removed. It was clear to Mr. B. that, after this operation, he would be sterile. Therefore, prior to the surgery, he had his sperm stored in the andrologic department of a dermatological clinic.[3] The storage capacity of this department was limited. Therefore, after two years, the department asked Mr. B. whether he was interested in continuing the storage. They informed him that they would discard the sperm within four weeks unless they were instructed otherwise. Within five days, Mr. B. responded that he still wanted his sperm stored. His letter was received at the department, but for reasons which remain unclear, it did not reach his file. Therefore, the laboratory was not informed and the sperm was discarded after an additional four months. Mr. B. was married in the same year. He learned that his sperm had been discarded only after he and his wife

H.A.M.J. ten Have and J.V.M. Welie (eds.), Ownership of the Human Body, 27–37.

decided to have children. He then sued the clinic for 25000 DM in damages. The lower court rejected the suit; the appellate court dismissed his appeal. The Federal Court, however, sided with him: the clinic was ordered to pay him 25000 DM in damages.

II. THE LEGAL POSITION UNDER GERMAN LAW

Mr. B.'s chance to have his own biological children had been destroyed. Under German law he had two options: he could demand a stated penalty against one person and/or he could bring a civil suit against the clinic. He had to renounce the first course of action because the likelihood of identifying a responsible person was minimal. Thus, only the option of filing a civil suit remained. Mr. B. had to sue for damages. The relevant paragraph of the civil code reads:

> Whosoever intentionally or by negligence wrongfully injures the life, the body, health, freedom or another law of another person is obligated to compensate for the resulting damage (Par. 823 BGB).[4]

Because the damage suffered by Mr. B. was not financial in nature, he could seek to obtain damages for his pain and suffering.[5] The pertinent paragraph reads:

> In case of bodily injury or injured health or in case of deprivation of liberty the injured may require an equitable compensation by money even when the damage is not an actual financial loss (Par. 847 BGB).[6]

Therefore, according to German law, damages are due if the body of a person has been culpably injured against that person's will. Monetary compensation may be awarded even when the damage is not financial.

III. THE ARGUMENTS OF THE FEDERAL COURT

It was not disputed by the parties, Mr. B. and the university which runs the clinic, that his sperm had been discarded culpably, namely, by negligence. But both lower Courts had refused to recognize the claim of bodily injury. It was their opinion that the cause of the psychosomatic disturbances alleged by the plaintiff was not an interference with the state of his bodily well-being (*Befindlichkeit*). Thus, if the plaintiff were to prevail, the Federal Court had to find that – in contrast with the view of the lower Courts – the body of the plaintiff had, indeed, been injured. The plaintiff had to show that his sperm, although separated from his body, in a legal sense belonged to his body.

The Court's reasoning has four steps. First, the Court confirmed a broad consensus among the legal profession that a part of the body which has been separated from the body is, legally, a thing and, therefore, is no longer part of a person. The right of the person with respect to his or her own body becomes a property right in the separated bodily part.

Secondly, the Court found that parts of the body which are destined to be returned to the body, in accordance with the wishes of the person affected, legally are not things. For instance, pieces of skin or bone may be temporarily separated from the body and then reunited with it as autografts. The same holds true for an autologous transfusion and an ovum which has been removed in order to be fertilized in vitro. These parts of the body continue to perform their natural function after reunification. Thus, if Par. 823 of the civil code protects the integrity of the human body and, thereby, protects the self-determination of the legal subject, the separated parts of the body continue to form a functional unity with the body and are thus also entitled to protection. To interfere with this unity against the wishes of the legal subject is bodily injury.

Thirdly, the Court investigated the situation in which bodily parts are permanently separated from the original body but united with a foreign body, i.e., the donation of blood and tissue. In this case the bodily parts are legally things because they no longer form a functional unity with the body of the donor and thus are not implicated in the donor's right to self-determination.

Finally, the Court addressed the circumstances of the present case. Sperm that was stored in order to perform a genuine bodily function, but simultaneously destined to be received by a foreign body represents a special case. I quote from the opinion:

> On the one hand the sperm is definitely separated from the body of the legal subject, on the other hand it is destined to fulfill a function typical of the body, procreation by the legal subject. If the stored sperm is to substitute for the lost capacity to procreate, it has....no less significance for the bodily integrity of the legal subject and for the personal self-determination and self-realization comprised in that integrity than the egg cell or other bodily parts which are protected by paragraphs 823 and 847 ... even after their removal from the body. Like the egg

cell removed for in vitro fertilisation and destined to be reimplanted, the stored sperm embodies for the legal subject the only possibility to exert his bodily functions to produce descendants to whom he passes on his genetic information.[7]

Thus, the Court regarded the interest of the female and the male providing them protection from any external interference with procreation, to be of the same kind and of equal value. Likewise, the protective intent of the legislation is held to be the same. Thus, the legal result should also be equal. The Court anticipated the following juridical objection: The stored sperm cannot enjoy the protection of bodily integrity, simply because they do not return to the body. But even if this were true, said the Court, the law would have to be applied as described. Paragraphs 823 and 847 would be expanded, but this expansion would be justified by the '*Persönlichkeitsrecht*', a right akin to the American right to privacy[8] of the legal subject, because his right to privacy is affected in the same way and to the same extent as the right to privacy of a female if one of her egg cells, destined to be reimplanted, were destroyed. Thus, the Court concluded: "Just as the female in that case, the plaintiff is lawfully entitled to damages on the basis of Par. 847 of the civil code".[9]

IV. FUNCTION AND ENDS

When the Court described the nature of the object to be protected by law, it quoted one of the famous commentaries: "The good to be protected by paragraph 823 is not the matter, but the field of being and determination of the person, which is given material form in the bodily state."[10]

This sentence is difficult to understand. Especially disturbing is the term 'matter/material form' – first mentioned in a negative and then in an affirmative sense. Therefore, one has to interpret the rest of the text. The key to comprehension, in my view, lies in the notion of 'function'. This term is used by the Court no less than four times: Bodily parts form, even during their separation from the body, a 'functional unit'; the right to self-determination makes the body and its separated parts still appear as a 'functional unit'; the sperm is destined to perform a 'function typical of the body'; this sperm is the only means left for Mr. B. to father a child or to perform his 'bodily functions to produce descendants'.

The Court takes it to be self-evident that a 'functional unit' enjoys protection by law. What is new is that legal protection is extended even to parts separated from the body. According to their view, the separated part belongs to the functional unit as well. Thus, the unity to be legally protected is not being shaped by spatial continuity (the surface of the body) but by function. Function refers even to the separated part of the body. The separated part has a function 'typical of the body', i.e., a function that occurs only in (living) bodies. An organ which has been removed from a dead body, e.g., during autopsy, would no longer be part of a 'functional unit' (that it is, nevertheless, not completely without legal protection is due to the fact that is had been part of a functional unit rather than that it is part of a corpse). What the Court calls 'function' is seen as a necessary condition of the body deserving protection inclusive of those parts that have been separated out. 'Function' also seems to provide the link between the notions 'field of being' and 'field of determination', because the functioning of the body belongs to its being as well as to its morphology.

What, then, does 'function' mean and why does it justify legal protection of the body? I shall first analyze colloquial language to show that when using the words 'function' or 'functioning' we include the idea of an end; then I shall distinguish two kinds of ends and, consequently, two kinds of functional units. Finally, I discuss what this means from the perspective of the law.

Functioning persons, things, mechanisms, and organs

The expression 'person A functions as p' means: Person A does something which preexists as a pattern of action (p) under a specified end, e.g., 'Mr. A functions as a referee, or a theater usher, or as a gate-keeper'. We may also say that Mr. A carries out the function of a referee, usher, or gate-keeper. What Mr. A carries out is a pattern of procedures or actions that can be carried out only by persons where its meaning preexists. In other words, meaning is not conferred on the procedure by the acting person, but the person finds that meaning in the social context and identifies with it. The specific course of events caused by Mr. A has the character of an action (or, of a bundle of actions connected by some meaning). But actions have ends that are contained in the preexisting meaning. In our examples, to be a referee means to care for fair play, to be an usher means to care for order, to be a gatekeeper means to keep out unwelcome people. The specific ends are contained in the tasks and defined by convention, or, respectively, by institutions, but not necessarily by the acting person herself.

The same linguistic relationship between course of events and effectuator of those events is found with things. The hammer 'functions as' tool, the diode 'functions as' indicator, a phenobarbitone 'functions as' an hypnotic. The ends of the respective things have been the causes for construing them; persons use them as means for those ends. The difference between things that serve a specific end, and a person who functions in a role is the fact that a person can identify with, that is, set those ends. Persons are able to deliberately set a course of events in motion and thereby produce an action. The action is a means to achieve an end the person has actually set. Things may even be machines or mechanisms, i.e., things that are construed to serve an end, and consist of parts which set each other in motion according to a preestablished plan. E.g., the chain of the bicycle 'functions as' power-transmission, the hard disk in a computer 'functions as' a data store.

The same linguistic relationship between course of events and effectuator of those events is found in (biological) organs. We can say 'the stomach serves digestion, the heart serves circulation, the kidneys serve excretion'; or, 'the stomach functions as a digestive organ, the heart functions as a pumping organ, the kidneys function as excretory organs'. Organs are neither persons nor inanimate things. They are neither able to set ends themselves nor is there a real person, convention, or institution which could have set ends for them. But, nevertheless, ordinary language treats them as having ends according to which they function.

Genuine ends, ends-as-if and natural ends

There is a significant difference between functions that are carried out by a person and functions that are carried out by a mechanism construed by a person and the function of an organ or a whole organism. In both cases we may rightfully speak of functions, but there are two different notions of 'end' involved. The philosophical tradition knows this difference under the terms 'constitutive' versus 'regulative idea of suitability',[11] or 'genuine end' versus 'end-as-if'[12].

Genuine end

"An end is an object of the choice (of a rational being), through the representation of which choice is determined to an action to bring this object about".[13] The 'rational being', i.e., the being capable of actions for which he can be held responsible sets ends by his power of representation, looks for means that are suitable to realize those ends and uses them for realization. Thus, the end presupposes a real subject having conscience and representations. It is true, the effort which is spent in looking for means varies considerably. Beings capable of acting can, e.g., find things in nature that prove to be suitable for certain ends already set, e.g., a stone of a certain shape for cutting, or deadly nightshade for making 'beautiful eyes'. Subjects capable of acting are able to produce means, e.g., the hammer, the diode, the phenobarbitone. They are even able to produce means that realize their ends 'by themselves', e.g., the watch. But the considerable differences of effort in looking for means do not detract from the fact that a real subject has set real ends (genuine ends).

End-as-if

What we find in living organisms and their organs are not genuine ends. We ascribe functions to organs although there is no real subject or consciousness that could have set those ends. Biological organisms merely behave – within some limits – as if somebody had set ends to them. We consider them, e.g., under the idea that self-preservation be their supreme end (which is impossible to seriously claim because of the lack of an end-setting subject). Self-preservation of living organisms is only an end-as-if, its suitability is only a 'regulative idea'. Empirically, we find in living organisms even details that cannot be interpreted as means that serve self-preservation, e.g., the appendix vermiformis in humans, or details that seem to be even unsuitable, e.g., the buoyancy which is given to penguins by their bird-like morphology.

Natural end ('Naturzweck')

On the other hand, the suitability (according to ends-as-if) which can be found in living organisms exceeds by far the suitability (according to genuine ends) invested in machines by human ingenuity. Living organisms not only maintain themselves in a status quo, and are – within certain limits – capable of repairing themselves, they even produce themselves and reproduce themselves. The first is called growth, the second procreation. Objects that contain this kind of suitability (to organize oneself under ends-as-if) were called by Kant natural ends ('*Naturzwecke*'):

> In such a product of nature any part, as it is present merely through all other parts, is conceived as existing on behalf of the others and of the whole, that is, as tool (organ); which, however, is not enough (because it could also be a tool of art and as such be represented as an end at all possible), but as an organ producing the other parts (consequently one another mutually), things like that being unable to be tools of art but only tools of nature which provides all material for tools (even for those of art); and only then and because of that such a product as an organised and itself organising being can be called a natural end.[14]

Referring to the difference between genuine ends and ends-as-if two types of functional units can be distinguished: functional units of the first type are arrangements of means that can be united under genuine ends; these are the machines and mechanisms used by end-setting subjects and cooperative social systems. Functional units of the second type are agglomerations of means that can be united under ends-as-if; these are the organs of organisms, the organisms themselves and the biological 'systems', e.g., species, ecological unit, animal or vegetable kingdom.

V. FUNCTION AND ACTION

Whenever the Court speaks of a functional unit, it refers to a unity of functions in the sense of a biological organism, i.e., to a functional unit of the second type. The human body is the sum of functioning organs; but it is not only a sum from which any other sum can be taken away. Insofar as bodily organs are present through each other, and, perceived under the end-as-if of self-preservation, form a whole organism no single organ or organ system can be removed without thereby affecting the whole organism.

But why should it be unlawful to remove something from that functional unit? This is the question that the Court had to answer. Interference with a functioning biological entity is not per se illegal, as is evidenced by the enormous degree of tolerated human interference with the kingdoms of animals and plants. The reason is the specific connection between the organism and the person. The human being as a living organism functions; the human being as a person acts. The human being as an organism is a natural end; the human being (a person) decides upon the ends of his other actions. But the human being as a person cannot separate him- or herself from his or her organism. Human beings are in all their actions and purposes dependent upon their bodies, even in thinking. This basic situation cannot be changed; it is descriptive of all human beings in the same way; it cannot be calibrated or traded among persons. If the law tolerated interference with one's body against one's will it would thereby tolerate a fundamental injustice. Thus, if the law is to protect the freedom of action of persons, it must protect persons' bodies, and this protection must then extend to all bodily functions which persons can use as means to achieve particular ends.[15]

This means that the implicit definition of the body that the German Court used is as follows. The body, in its legal sense, is the sum of those organs and functions of an organism of a person that a person can use as a means (of achieving his or her chosen ends).

Klinikum der Philipps-Universität
Marburg, Germany

NOTES

1 *Bundesgerichtshof*, Urteil 9. November 1993, Aktenzeichen VI ZR 62/93.

2 The author expresses his gratitude to Arlene Judith Klotzko, J.D., who provided helpful suggestions on an earlier draft.

3 A department specializing in disorders relating to the male sex.

4 "Wer vorsätzlich oder fahrlässig das Leben, den Körper, die Gesundheit, die Freiheit, das Eigentum oder ein sonstiges Recht eines anderen widerrechtlich verletzt, ist dem anderen zum Ersatze des daraus entstehenden Schadens verpflichtet." (Par. 823, Abs.1 Bürgerliches Gesetzbuch) [6].

5 According to German law, violation of contract is a sufficient reason for compensatory damages, but not for damages for pain and suffering.

6 German: "Im Falle der Verletzung des Körpers oder der Gesundheit sowie im Falle der Freiheitsentziehung kann der Verletzte auch wegen des Schadens, der nicht Vermögensschaden ist, eine billige Entschädigung in Geld verlangen." (Par. 847, Bürgerliches Gesetzbuch) [6].

7 German: "Einerseits ist das Sperma endgültig vom Körper des Rechtsträgers getrennt, andererseits ist es dazu bestimmt, eine körpertypische Funktion, die der Fortpflanzung des

Rechtsträgers, zu erfüllen. Jedenfalls wenn, wie hier, die Spermakonserve die verlorene Fortpflanzungsfähigkeit substituieren soll, hat sie für die körperliche Integrität des Rechtsträgers und die in dieser begriffene personale Selbstbestimmung und Selbstverwirklichung ... keine geringere Bedeutung als die Eizelle oder andere körperliche Bestandteile, die nach dem zuvor Gesagten auch nach ihrer Entnahme aus dem Körper von dessen Schutz durch die Par. 823 Abs.1, 847 Abs. 1 BGB erfaßt sind. Ebenso wie die zur Befruchtung entnommene und zur Reimplantation bestimmte Eizelle verkörpert das konservierte Sperma für seinen Rechtsträger die im Streitfall einzige Möglichkeit, seine körperlichen Funktionen zur Hervorbringung von Nachkommen, denen er seine genetischen Erbinformationen weitergibt, zur Geltung zu bringen." *Urteil des Bundesgerichtshofs*, 9. Nov. 1993, Aktenzeichen VI ZR 62/93.

8 The German term 'Persönlichkeitsrecht' means the right to freely express one's personality, which is a 'Grundrecht', granted by the Constitution.

9 German: "Ebenso wie der betroffenen Frau in jenem Fall, so steht dem Kläger hier auf der Grundlage des Par. 847 BGB ein Anspruch auf Schmerzensgeld zu." *Urteil des Bundesgerichtshofs*, 9. Nov. 1993, Aktenzeichen VI ZR 62/93.

10 German: "Schutzgut des Par. 823 Abs. 1 BGB ist nicht die Materie, sondern das Seins- und Bestimmungsfeld der Persönlichkeit, das in der körperlichen Befindlichkeit materialisiert ist". *Reichsgerichtsräte-Kommentar zum Bürgerlichen Gesetzbuch*, 12. Aufl., Par. 823, Randnummer 9 [7].

11 German: Konstitutive versus regulative Idee der Zeckmäßigkeit; Immanuel Kant, *Kritik der Urteilskraft*, Par. 62-78 [4].

12 German: Echter Zweck versus Als-ob-Zweck; N. Hartmann, *Teleologisches Denken*, Kap. 7-9 [1].

13 German: "Zweck ist ein Gegenstand der Willkür (eines vernünftigen Wesens), durch dessen Vorstellung diese zu einer Handlung, diesen Gegenstand hervorzubringen, bestimmt wird". Immanuel Kant, *Metaphysik der Sitten, Einleitung zur Tugendlehre*. Akademieausgabe Bd. 6, S. 381. [3]. English translation by Mary Gregor, *The metaphysics of morals*, p. 186 [5].

14 My translation, stresses by Kant. German: "In einem solchen Produkte der Natur wird ein jeder Teil, so wie er nur durch alle übrigen da ist, auch als um der anderen und des Ganzen willen existierend, d. i. als Werkzeug (Organ) gedacht; welches aber nicht genug ist (denn er könnte auch Werkzeug der Kunst sein und so nur als Zweck überhaupt möglich vorgestellt werden), sondern als ein die anderen Teile (folglich jeder den anderen wechselseitig) hervorbringendes Organ, dergleichen kein Werkzeug der Kunst, sondern nur der allen Stoff zu Werkzeugen (selbst denen der Kunst) liefernden Natur sein kann; und nur dann und darum wird ein solches Produkt als organisiertes und sich selbst organisierendes Wesen ein Naturzweck genannt werden können". Kant, *Kritik der Urteilskraft*, Par. 65, Akademieausgabe Bd. V, S. 281-295.

15 A more detailed discussion of the different bodily functions to be protected by law and the scope and moral justification of the law with respect to the human body can be found in Heubel und Freund [2].

BIBLIOGRAPHY

1. Hartmann, N.: 1951, *Teleologisches Denken*. Walter de Gruyter, Berlin, Germany.

2. Heubel, F., Freund, G.: 1996, 'Vernichtetes Sperma als Körperverletzung. Eine rechtsethische Diskussion des Körperbegriffs', *Zeitschrift für medizinische Ethik* 42, 129-141.
3. Kant, I.: 1966, *Metaphysik der Sitten*. Felix Meiner, Hamburg.
4. Kant, I.: 1959, *Kritik der Urteilskraft*. Felix Meiner, Hamburg.
5. Kant, I.: 1991, *The metaphysics of morals* (transl. Mary Gregor). Cambridge University Press, Cambridge, UK/New York/Melbourne.
6. Palandt, O. (ed.): 1994, *Bürgerliches Gesetzbuch 53*. Aufl. C. H. Beck, München.
7. RGRK: 1974, *Das Bürgerliche Gesetzbuch mit besonderer Berücksichtigung des Reichsgerichts und des Bundesgerichtshofs, herausgegeben von Mitgliedern des Bundesgerichtshofs*. 12. Aufl. Walter de Gruyter, Berlin/New York.

HUB A.E. ZWART

WHY SHOULD REMUNERATED BLOOD DONATION BE UNETHICAL?

Ethical Reflections on Current Blood Donation Policies and Their Philosophical Origins

I. INTRODUCTION

As any survey of the literature on blood donation will reveal, most experts have been advocating unremunerated donation, not only for ethical reasons, but also for reasons of safety and efficiency. Yet, a considerable number of countries have to rely one way or another on paid donors. They either have a commercial system of their own, or import commercial blood products from other countries. The commercialization of blood was reinforced by the development of new techniques such as intensive plasmapheresis. The over-all demand for blood has continued to increase since then. Experts agree that we are in need of an international blood policy that is both ethical and realistic. Even in the case of unremunerated donations, the maintenance of an adequate blood supply will partly have to rely on market transactions. How far ought we to allow commercial mechanisms to penetrate this realm of altruism?

In 1981, Piet J. Hagen published *Blood: Gift or Merchandise*, an extensive study into the possibilities and difficulties of developing an international blood policy [6]. The title contains a clear statement of the dilemma, which is only partly ethical. Other aspects, such as efficiency, are involved as well. As to the ethical aspects, however, Hagen maintains that, although he prefers nonremunerated donation of blood, the payment for blood cannot in itself be considered unethical ([6], p. 186). If a commercial blood bank or plasma center complies with all the regulations, if a donor gives his or her informed consent, and if he or she is not compelled to donate by whatever circumstances, the paid donation is a transaction that is not morally inferior to other transactions in the market. Hagen realizes, however, that payment for blood *may* have consequences that some consider unethical. For instance, it may produce inequality between poor and rich countries, or facilitate the payment for

H.A.M.J. ten Have and J.V.M. Welie (eds.), Ownership of the Human Body, 39–48.

other body parts, and eventually lead to the commercialization of the human body itself.

I recognise the importance of safety procedures, such as informed consent, and the absence of force and pressure. As obvious as the voluntary character of donation might seem, we are told that in Beijing, China, for instance, there is a law that renders blood donation obligatory for all healthy adults below the age of 55 [1]. Yet, I think that other moral aspects should be considered as well. My contribution, therefore, is aimed at broadening the scope of the ethical debate on whether remunerated blood donation should be considered immoral. In order to attain a full moral evaluation of blood policies, our understanding of basic notions such as personhood and ownership have to be taken into account. It is my contention that in the ethics literature on blood donation two different moral vocabularies are encountered. These vocabularies are embedded in incompatible metaphysical and anthropological views on ownership and personhood, and on the relationship between person and body.

In the first section I will briefly review the ethical debate on paid blood donation, building on earlier publications [13]. Next, I will emphasize that moral reflection is never neutral, but always implies a substantial view on human existence. In the ethical debate on blood donation, two different moral vocabularies emerge, the gift vocabulary and the commodity vocabulary. These vocabularies are embedded in two incompatible perspectives on ownership and personhood, which I will refer to as the modern liberal and the premodern communitarian perspectives. Whether we consider a particular ethical argument for or against paid donation convincing, will depend on our responsiveness to the basic views on ownership and personhood from which it was derived. Finally, I will conclude that, although the modern liberal perspective allows us to capture important aspects of moral experience, it does not contain the whole story and should be supplemented by the gift vocabulary. Otherwise, our capacity to give voice to our moral repugnance over the commercialization of the human body is bound to decline.

II. ARGUMENTS IN FAVOR OF AN UNPAID SYSTEM

In *The Gift Relationship* Richard Titmuss compared systems of collecting and distributing human blood for therapeutic use in Great Britain and the United States [11]. He found the British system, which utilized blood collected from unpaid donors only, to be far superior in a number of respects to the American approach, which used a combination of commercially and non-commercially supplied blood. Titmuss not only considered the British system more efficient, he also claimed that blood from commercial donors carried greater risk of transmission of, especially, hepatitis, than blood from unpaid donors. In the commercial system, donors are more likely to be found among alcoholics, drug addicts, and other groups with comparably great health risks. Among the causes to which the difference between both systems, both in terms of safety and in terms of efficiency, should be attributed, Titmuss notably mentioned the very presence of a commercial market for blood in the United States. Much of the fascination of Titmuss's book lay in his exposition of the special significance of the 'gift' of blood. Others, however, have stressed the inability of a non-commercial blood sector to supply the entire medical demand for blood. Moreover, they claimed that a non-commercial system deprives individuals of the freedom to sell their blood.

Singer [9,10] agrees with Titmuss in many respects. When commercial and non-commercial systems for obtaining blood exist side by side, he claims, voluntary donors are discouraged to donate for altruistic reasons. The availability of blood as a commodity affects the nature of the gift that is made if blood is donated. If blood cannot be bought, giving blood remains something special, an act of providing for strangers, without receiving a reward. The claim brought forward by some, that the prohibition of a market for blood products impedes the individual's freedom to sell his blood, is refuted by the argument that to take away the prohibition would lead to a decline in freedom as well. What allowing a market in blood prevents, is the freedom to donate something that cannot be bought. To accept one person's right to sell his or her blood is to deny the rights of others to give something that is, literally, priceless. Moreover, to accept the right to sell is to deny to the community the right to shape its institutions so as best to encourage altruism among its members. In countries like France, Belgium, and the Netherlands, all the whole blood required for medical purposes comes from unpaid donors.

The donors receive no reward, beyond "the knowledge that they have given something above price" ([10, p. 49). Many individuals feel moral repugnance when they are confronted with the idea of a market for blood, where blood and blood products are viewed as commodities. According to many, the most appropriate model for thinking about blood donation is the idea of the gift.

However, it may be too simplistic to view the gift relationship exclusively as good, and a commercialized system exclusively as bad. In premodern times the gift relationship played a prominent role in public life, and developments toward more business-like human relationships have freed us from the enslavement of gift relationships [7]. Still, Murray goes on, the gift relationship does play a positive role in social life, because it promotes solidarity in the face of powerful forces of alienation, and serves essential social values that are not well served by markets, commerce and contract. Mass bureaucratic societies need to affirm what it is citizens share with their neighbours. Gifts, and especially gifts of the body are one of the most significant means we have to affirm that solidarity. Although the commercial market system is indispensable for behavior regulation on a large scale, other means should be used as well to preserve some kind of community in the face of the damage to human solidarity that is caused by the market system.

In the Netherlands, Gevers [5] made a noteworthy contribution which, apart from legal considerations, contained ethical considerations as well. Gevers claims that, due to technological developments, parts of the human body, such as blood, tend to gain a more independent status. This inevitably raises the question whether it is legally or morally permissible for parts of the human body to be subjected to commercial exploitation. According to Gevers, it is still generally accepted that the donation of body material ought not to be commercialized. Donorship is expected to be prompted by higher motives, and remuneration would offer an improper incentive to donate (apart from the fact that non-commercially supplied blood is considered more safe). Nevertheless, Gevers points to the fact that some experts have been advocating the commercialization of blood donation. Furthermore, in view of the growing independency of human body material, it is possible that the willingness to donate without remuneration will decline in the future. Gevers, however, maintains that commercialization should be rejected. The prospect of a financial reward for blood donation is bound to harm the intrinsic value of the human body.

Apart from these ethical considerations, unpaid donation is favored by many for reasons of safety and availability. Although arguments concerning the safety and scarcity of blood supplies are technical arguments, it goes without saying that they are morally relevant, because considerations pertaining to the patient's safety and to the availability of scarce resources clearly entail moral issues. Furthermore, it goes without saying that the ethical weight of the risk argument has increased dramatically since HIV can be transmitted by donated blood. Most of the current newspaper headlines on blood donation address the safety issue.

The moral considerations that emerge in this survey evoke some important ethical questions. Why should blood donation be a case for altruistic rather than benefit-seeking behavior? Why should it be more appropriate to think about blood donation as a gift, rather than as a commodity? Why should individuals be free to donate their blood, but not to sell it? Why should we feel moral repugnance when confronted with the idea of a market for blood? What is meant by the intrinsic value of the body, harmed by the prospect of a financial reward for blood? In the next section I indicate that moral reflection can never be neutral. Afterwards, I will identify the basic views that have informed the debate on blood donation.

III. BIOETHICS AND MORAL TRADITIONS

In contemporary bioethics the conviction has emerged that moral dilemmas can and should be discussed in a *neutral* moral language [4]. A language, that is, that does not rely on any particular moral tradition, and refrains from appealing to substantial views on human existence, whether they are of a metaphysical or of a religious nature. In my view, however, moral languages can never be neutral. They always tacitly contain some broader views. The very notion of consent, for instance, is a substantial moral notion that belongs to a particular tradition (modern liberalism). It was introduced by Locke in his famous *Second Treatise on Government*. Although I do not deny that the notion of consent captures an important aspect of moral existence, it should not be cut away from the particular view on human life, bodily existence, and personhood in which it was originally embedded.

In Locke's *Second Treatise* we find the notion of consent situated in a conceptual network, together with other liberal notions and concerns. In

Locke's view, consent is a conceptual tool that allows for the regulation of social exchange, for the maintenance of peaceful coexistence among competitive individuals. And here property comes in, for it is as the owners of personal property that we participate in social exchange. This means that the notion of consent, of voluntary participation without force, inevitably points to other basic concerns, such as ownership and personhood. It means that the *ethic* of consent eventually points to anthropological considerations concerning persons, bodies, and things. Although the capacity of the modern liberal perspective to capture important aspects of moral experience is beyond doubt, it is one possible perspective among others. In some respects, these other perspectives might prove more congenial. In the debate on blood donation, the modern liberal perspective has to compete with an older moral vocabulary, embedded in a different perspective on human existence.

IV. OWNERSHIP AND PERSONHOOD: A TALE OF TWO PERSPECTIVES

As any survey of the history of philosophy will reveal, the modern liberal view on personhood and ownership is more recent and less obvious than it might seem. The modern liberal way of speaking and thinking about property is one possibility among others. According to Thomas Aquinas, for instance, we are not the owners of our property as it is understood today [2]. Rather, the things we 'own' are entrusted to our care and stewardship. Reading Saint Thomas reminds us that contemporary notions of ownership seem rather at odds with the idea of creation. This incompatibility notably applies to the body. According to the moral tradition represented by Thomas Aquinas, the human body is a gift entrusted to our care rather than a thing we have at our disposal, or even a kind of merchandise we own and are allowed to sell.

In many realms of human existence, the deferential view on ownership encountered in the writings of Thomas Aquinas and others, gave way to the less awe-inspired, liberal approach. Man has come to consider himself as the owner, rather than merely as the guardian of his property, with the possible exception of the realm of bodily existence. For centuries, this has remained a stronghold for the intuition that self-determination is in itself restricted, an intuition, however, which could not be adequately voiced in terms of the modern liberal commodity vocabulary. The debate on commercial blood policies indicates that now the struggle between moral perspectives has reached the human body itself. Can blood be considered a thing we own (or even a merchandise we sell), or is it rather a gift entrusted to our care? The last option would entail that bodily parts can be donated, but not sold to others.

In many respects, it is not astonishing that blood should become the 'corpus delicti' by means of which the modern liberal view on ownership tries to obtain access to the human body. Unlike kidneys and other organs it is a renewable resource. More important, however, is the fact that the modern liberal view has been supported and prepared by the emergence of the modern life sciences, especially physiology. Although physiology is said to be the science that studies life, or rather the difference between life and death, its initial source of knowledge consisted of corpses. Physiology could only prosper after the deferential aversion to the dissection of corpses had declined. A particular, less awe-inspired way of looking at and thinking about the body was made possible by this event. And the first thing to be discovered by this invasive movement into the body, was blood circulation. There resided the difference between life and death. It contained the body's infrastructure that allowed the other organs to function and interact. The ethical debate seems to follow the same track initially laid bare by the founding physiologists. From here, the body can be further exposed to commercialization.

Someone very much engaged in the dissection of animal bodies at that time, was the philosopher Descartes [3]. His main effort, however, did not consist in any physiological discovery, but rather in developing the kind of metaphysics this newly emerging attitude towards the body called for. According to Descartes, two kinds of substances can be encountered in the world: 'extended things' (like the human body) and 'thinking things' (like the human mind). Such a view inevitably turns the human body into a complicated machine at the service of the human mind. Descartes' metaphysics provided the epistemological space in which modern liberal conceptions of ownership and personhood could emerge. It allowed for a rigid distinction between the person (soul) and his body. Personhood became strongly connected with the subjective and cognitive aspects of human existence. Conceptions of personhood, grounded in such metaphysics and implying a moral degradation of the physical aspects of the human being, have become self-evident to such a degree that one might find it difficult to imagine that things used to be quite otherwise. According to Thomas Aquinas, personhood refers not merely to the subjective and cognitive aspects of the human being, but rather to the human being as an integrated whole, to the *complete* human being, encompassing both body and soul. In such a view, the body simply cannot be conceived of as something that is at a person's disposal, for it is itself integral to the person as such. This perspective on personhood prevents the emergence of a commodity vocabulary in talking about the human body. It rather produces a gift vocabulary.

These premodern views on ownership and personhood have lost much ground to the modern perspective. Yet, some aspects of it are still audible in the writings of theologians and philosophers, but also in ecclesiastical teaching. In 1990, for instance, when the present Pope addressed the subject of voluntary donation, he pointed to the analogy between an individual's gift of a part of his body to another human being, and the gift bestowed on all of us by Jesus Christ Himself. It is a gift that overcomes death. The gift of blood serves as an antidote against our society becoming excessively market-oriented. As we received our body as a gift we can give to others. Not as a transaction but as a gift without remuneration. We do so, not as profit-seeking individuals, but as members as a human community. Our sense of community is restored by these acts of giving [8,12].

V. DISCUSSION

If we remain within the boundaries of the modern liberal perspective, remunerated blood donation *as such* will not be considered immoral, and it will be impossible to make sense of the moral repugnance evoked in some by the idea of a market for blood. Such repugnance points to aspects of moral experience that cannot adequately be voiced in terms of the vocabulary of commodity and consent. The very notion of a gift is a survival of an older, more deferential perspective which, after loosing ground to modern liberalism in many realms of human existence, now seems to lose its hold on the human body as well. I have already contended that the modern liberal perspective allows us to capture important aspects of moral experience. It gives voice to our repugnance over the use of human beings without their voluntary consent, as is the case in the Beijing legislature on blood donation noticed above. But could it be the whole story? Or should we rather grant the gift vocabulary a right to speak as well? Is there still a reason for thinking in terms of such a premodern, deferential perspective at all?

In speaking about moral experience, I did not yet consider the actual motives of blood donors. The absence of remuneration does not imply that all of them will view the act of donating in terms of the gift vocabulary. Other, more prudential and less altruistic motives might be involved as well. One of the non-altruistic reasons to donate might be that donorship implies a regular medical check-up free of charge. Or the donor might anticipate that sooner or later, he himself will be a patient in need of blood. Being a donor, he will then feel entitled to reciprocity. And this is quite unlike the notion of a gift, which does not entail the prospect of being repaid. This, however, does not remove the conviction that something is lost should we allow the modern liberal perspective to establish an absolute monarchy. Some aspects of donation will become distorted if we limit our moral vocabulary. Both on the moral and on the societal level, we are in need of a counter force, a counter vocabulary, that allows us to articulate aspects of moral experience concealed by modern liberalism. The predominance of this one perspective would pose a threat to moral pluralism. In the ethical debate on blood donation, both moral vocabularies should remain audible, in order to preserve an adequate scope for the debate. Otherwise, we would be left without a congenial vocabulary when repugnant responses against commercialization become too strong.

Center for Ethics
Catholic University of Nijmegen
Nijmegen, The Netherlands

BIBLIOGRAPHY

1. Anonymus: 1992, 'China: blood donation by order', *The Lancet* **339**, 545.
2. Aquinas, Th.: *Summa Theologica*, IIa IIae, qq 65-66.
3. Descartes, R: 1953, *Traité de l'homme*, in Oeuvres de Descartes, Colbert, Paris, pp. 803-873.
4. Engelhardt, H.T.: 1986, *The Foundations of Bioethics*, Oxford University Press, New York, NY/Oxford, UK.
5. Gevers, J.K.M.: 1990, *Beschikken over cellen en weefsels*. Kluwer, Deventer.
6. Hagen, P.J.: 1981, *Blood: Gift or Merchandise. Towards an international blood policy*, Alan R. Liss, New York, NY.
7. Murray, T.: 1987, 'Gifts of the body and the needs of strangers', *Hastings Center Report* **17**, 30-38.

8. Anonymous: 1991, 'Pope to group on organ transplants: New way of sharing life with others', *Osservatore Romano* (English Edition). June 24.
9. Singer, P.: 1973, 'Altruism and commerce: A defense of Titmus against Arrow', *Philosophy and Public Affairs* **2**, 312-320.
10. Singer, P.: 1983, 'The blood feud: Round two', *Hastings Center Report* **13**, 48-50.
11. Titmuss. R.M.: 1971, *The Gift Relationship: from Human Blood to Social Policy.* Allen & Unwin, London, UK.
12. Vosman, F.: 1992, 'Het broze lichaam', in V. Kirkels (ed.), *Transplantatie en mensbeeld.* Baarn, Ambo, pp. 89-104.
13. Zwart, H.: 1993, 'To be paid or not to be paid. Ethical aspects of the tension between unremunerated, self-sufficient blood donation and the common European market'. *International Journal of Bioethics* **4**, 131-136.

WIM J.M. DEKKERS
HENK A.M.J. TEN HAVE

BIOMEDICAL RESEARCH WITH HUMAN BODY "PARTS"

I. INTRODUCTION

For several decades, the morality of medical research with human subjects has been one of the main areas of activity and reflection in philosophy of medicine and bioethics. It is argued that precisely the issues and problems of medical experimentation have led to the emergence of modern bioethics [18]. The body of literature that has developed since then focuses exclusively on experimentation with human subjects 'as a whole', that is, persons and their (entire) bodies [15,19]. Much less attention has been given to biomedical research (medical experimentation) with human body parts. For example, the Declaration of Helsinki contains no ethical guidelines concerning the collection, storage, and use of human body parts. However, in present-day medicine, human tissues and body fluids are increasingly used for research purposes. The availability of human body parts is of undeniable importance for basic research, for research aimed at improving therapies or developing new treatments. The ethical aspects of procurement, donation, and allocation of human body parts have been discussed in the context of blood transfusion and transplantation medicine, but the scope for using human body parts in the treatment of other patients is rapidly expanding. Human tissues also play an important part in quality control in the health care system. Finally, they are sources for the manufacture of diagnostic and therapeutic aids. These various uses of human body parts are connected with various practices of biomedical research. In some cases, body parts are directly obtained for research purposes, but in other cases human tissues, organs, and body fluids are used for purposes other than those for which they were originally obtained. These practices raise questions concerning the ethical and legal issues involved [12], but also lead to an exploration of the nature and value of human bodies and body parts. In this chapter, we shall focus on some of the ethical problems related to biomedical research with human body parts.

H.A.M.J. ten Have and J.V.M. Welie (eds.), Ownership of the Human Body, 49–63.

II. HUMAN BODY PARTS

In exploring the value and use of human body parts, it is important to make a conceptual distinction: when an organ is fully integrated within the organism as a whole, it is a *part* in the strict functional sense; when it is extricated or separated from the organism's whole, it is a body element. The next question that arises is what kind of entities are to be regarded as human body parts? What perspective and which principles do we use to distinguish true functional parts of the human body; or, how do we divide the body as a whole into its separate parts? Two approaches seem possible.

From a *synthetic* point of view we can distinguish between a number of 'external' and 'internal' body parts according to the structural and functional classification systems of human anatomy and physiology. For example, there is a structure of body parts connected and articulated by the skeletal joints: arms, legs, hands, feet, fingers. Within a *functional* structure, various anatomical parts may be identified and distinguished. Bodies are anatomized into external parts which can be considered as functional units or building blocks within a larger bodily segment. From this macro-perspective view we may recognize a head, a chest, breasts, ears, noses [11]. As internal body parts we may describe the organs which are 'inside' the body and which have specific functions.

From an *analytic* approach the human body may be seen as a complex structure consisting not only of organs, but also of tissues, cells, and subcellular structures. Our body contains, for instance, 206 bones and more than 600 muscles. The body of an adult person of circa 150 lbs consists of approximately 60 billion cells, differentiated in 200 cell types; it also has an extracellular matrix which can be fibrous, jelly-like, or liquid [14]. From this analytic perspective it is useful schematically to distinguish between solid and liquid body parts and between parts that are (normally) 'inside' the body, and parts that are discharged, produced, secreted and excreted by the body. This twofold distinction leads to the following scheme:

(1) solid parts, normally in the body (e.g., organs, egg-cells);
(2) body fluids, normally in the body (e.g., blood, lymph, lumbar fluid);
(3) solid parts discharged by the body (e.g., hair, nails, feces, superficial cells of the skin);
(4) body fluids discharged by the body (e.g., urine, sweat, tears, milk, sperm).

Within this scheme there are borderline cases. Some body parts can not be subsumed under one of the four categories mentioned, for example: the placenta, bile, sputum, and pus. Fetal tissue obtained through an abortion is another example of a borderline case.

Leaving this classification problem for the moment, all body materials, body fluids, and body products mentioned may be regarded as human body parts in which medical scientists are interested. What, then, should the definition of human body parts be? It is our suggestion that we define 'human body parts' in the strict sense as follows: human body parts consist of organic material which is or used to be an *integrated* part of the human body, and has its origin in or is produced by the human body. This definition takes two characteristics as significant.

1. *Integration of the part within the whole of the human body.* Bullets remaining in the body of a surviving soldier or surgical material accidentally left within the abdomen of a patient do not become body parts. To be a body part in the strict sense, an entity must be part of the body, in the sense that it is or has become an integrated component of the body as a whole. For this process of integration, it also seems that the external boundaries of the body are highly significant. There must be some kind of 'internalization' of the entity involved in order to identify or transform it into a body part. For example, the glasses of a myopic person are not considered a body part, while the implanted kidney of a person with end-stage renal disease is definitely a body part. On the other hand, integration, although it is a necessary criterion for an entity to be a body part, is not a sufficient criterion. A second characteristic is essential.

2. *Origin of the part in the human body.* In order to be a human body part, the part must either be produced by a human body or have its origin in it. The producing or originating body, however, does not need to be the body wherein the body parts are *integrated* at any particular moment. A transplanted kidney (element) can 'move' from one body to another, but integrated into the donor or the recipient it is a body *part*. A xenograft, on the other hand, can not be, according to our definition, a human body part, even when it is functioning within that body. From our definition it also follows that implanted artificial materials – artificial joints, heart valves, or lenses – are not considered body parts.

In the scientific literature it is usually taken for granted what a human body part is. Definitions, if at all given, are vague and implicit. Dworkin and Kennedy in discussing the rights which may exist over the whole or parts of the human body, refer to the terms 'human tissue' and 'biological material', which are in fact used as synonyms for body parts. They define these terms as follows: "every aspect of a person's being, ranging from body waste (such as urine, feces, hair, nail clippings), to a list representing an atlas of the human body" ([7], p. 291). Munzer, in an extensive discussion of the problem of property rights in body parts, presents the following definition of body parts: "any organs, tissues, fluids, cells, or genetic material on the contours of or within the human body, or removed from it, except for waste products such as urine and feces" ([16], p. 259). Thus it is obvious that the same entity (for example, urine) can be a body part according to one definition, but not according to the other. Indeed, integration and origination within the human body are essential characteristics of being a human body part in the strict sense.

III. BIOMEDICAL RESEARCH

There are many different purposes for which body material is obtained and many reasons for medicine to be interested in human body parts: diagnostic, therapeutic, scientific, educational, commercial. In current practice, human tissue samples are more commonly stored than destroyed; more often than not, patients and donors are not aware that material taken from them is stored. A committee of the Health Council of the Netherlands, exploring the current practice of the collection, storage, and use of human tissues, found that the use of human tissue for purposes other than that for which it was originally taken is so widespread in practice that it can be called normal ([12], pp. 61-67). If well preserved, body parts can survive the (whole) body itself for many years, as illustrated in anatomical and pathological museums. Almost all body parts can be used in one or another way as material on which a medical diagnosis can be built: blood, urine, a liver biopsy, etc. Many body parts can be used to produce drugs (human chorion gonadotrophin in the urine of menopausal women), cosmetic products (placenta, hair for a toupee), or diagnostic tests. The list of what can be done with human body parts is increasing day by day [12]. Recently it has been discovered that the blood of the umbilical cord of fetuses or neonates is a potential source of therapeutic substances. The use of umbilical blood cells has great advantages as an alternative to bone-marrow transplantation in patients with leukaemia [2]. Following earlier initiatives to built up biological dossiers through creating tissue banks and cell banks, there are now umbilical cord blood banks. The more medical science makes progress, the more it is interested in new and smaller body parts. An outstanding example of this development, of course, is modern molecular genetics. The geneticist only needs some small amount of DNA to carry out his or her research; from minuscule amounts of body material, a wealth of information can be generated.

The various uses of human body parts in the treatment of patients, in quality assurance of laboratory findings, in teaching, or in preparation of diagnostic and therapeutic products, all presuppose use of body parts in research. In practice, the distinction between the use of human body parts in research and their use in the application of the results of research is not sharp. Not only is the use for other than research purposes almost always preceded by medical research, but also fundamental as well as applied research rely on the use of human tissues, for example, in applying quality assurance techniques using serum or tissue samples as reference material.

From a legal point of view, biomedical research with body parts is more often than not included in definitions of experimentation with human beings. For example, in the Netherlands the recent legislation "Regulations on Medical Research involving Human Subjects" (Medical Research involving Human Subjects Act) uses the following broad definition of an experiment: medical-scientific research which involves subjecting persons to certain procedures or imposing particular behavioral requirements.[1] Following this definition, strictly interpreted, it might be argued that donating particular body parts (hair, nails, urine) for experimental research is not covered by the Act, except in those cases requesting a (temporary) change in the normal behavior of subjects. A crucial question then is, whether simple acts such as collecting urine may be seen as 'imposing on persons a particular behavioral requirement'. However, the collection and use of human tissue as part of an experiment is generally regarded as covered by the Act. On the other hand, the Act does not apply to the use of body parts that remain after an experiment is concluded. Arguing from the principle that an individual has the right to determine what will happen to his or her body (or its parts), informed consent is considered an absolute requirement for collecting and using body parts for research purposes. When, however, body parts remaining after the original research project has ended are used for other purposes, including new research projects, the same requirement is not explicitly stated in the Act. It is argued that the storage and use of tissues after the completion of an experiment need the informed consent of the subject [12]. When it is clear in advance that the collected tissues are also of value in future research projects, consent should be obtained at that time; otherwise subjects may assume that the tissues will be destroyed as soon as the original experiment is completed. When stored tissues are used for purposes not known at the time of collection, it depends on the identifiability of the material how the patient or donor will be involved. The Health Council committee, mentioned earlier, proposes that at the earliest possible stage express consent must be obtained for any further use of human body parts which are directly or indirectly identifiable. When the material stored or used is not identifiable, patients or donors should be informed and given the opportunity to register an objection. However, the committee acknowledges that it is as yet not common practice to seek consent for the storage and further use of human body parts; it also recognizes that the circumstances of individual cases and projects should be taken into account ([12], p. 73).

Epidemiological studies raise somewhat different ethical issues than those that usually confront IRBs in their review of biomedical research. The use of stored body parts (elements) is extremely valuable for epidemiological research [5]. Old tissue samples taken from individual patients can generate information on the incidence, prevalence, and distribution of diseases. For example, a study in 1990, examining with the latest techniques stored tissue specimens from a Manchester patient who died of a mysterious disease in 1959, showed that the Aids virus was present at that time [6].

IV. RESEARCH AND BODY PARTS

In this section we discuss examples of biomedical research with human body parts, as recently reviewed by an IRB. We offer a short description of each research proposal followed by an explication of the most important ethical problems. In the next section we return to the ethical aspects inherent in these cases.

Case A: Intravenous human immunoglobulin treatment of children with intractable childhood epilepsy

In patients with childhood epilepsy several immunological abnormalities have been reported. Epilepsy is sometimes associated with a selective immunoglobulin A deficiency. The research proposal is a randomized, double-blind, controlled study of the efficacy of a particular immunoglobulin in the treatment of some forms of childhood epilepsy, based on the clinical and experimental data suggesting an autoimmune pathogenesis of epilepsy.

The drug concerned is a lyophilized, virus inactivated concentrate of functionally intact aggregate-free immune globulin. It is manufactured from pooled human plasma. Plasma units are obtained from licensed plasmapheresis centers in Europe and the United States. The plasma is tested with regard to the presence of antibodies of different viruses (HIV, hepatitis, etc). Potential viral contaminants are inactivated and/or eliminated during the manufacturing process.

Case B: Markers for tumor progression and metastasis

Metastasis of tumor cells is a multi-phase process. One of the phases requires the invasion of the tumor cells in the basal membrane. This process can be studied *in vitro* by bringing tumor cells into contact with a basal membrane and measuring their invasive behavior. The aim of this study is to examine the invasion of human melanoma cells in a basal membrane isolated from normal placental tissue. Each normal placenta can be used for this purpose.

The application for approval of this study included a typical example of the researchers' misunderstanding that placentas may be used for this purpose because they are otherwise discarded after childbirth. The researchers wanted to use placentas without the informed consent of the women concerned.

Case C: Clinical and molecular aspects of human genomic imprinting

The term 'genomic imprinting' refers to a selective inactivation of paternal or maternal genes on a particular chromosome. This mechanism may lead to 'uniparental disomia', in which two homologue chromosomes of one and the same parent come together in one individual. This, in turn, may lead to congenital, non-hereditary, malformations.

For the study of uniparental disomia, the researchers need blood of patients with a particular congenital malformation and blood of their parents as well. From this blood lymfoblastoid cell lines, which may serve as material for DNA isolation, will be manufactured. For the study of genomic imprinting fetal tissue and maternal blood are needed. Physically the research is hardly burdensome to the woman since only a small amount of blood is needed.

For the study of genomic imprinting the researchers wanted to use material of fetuses, after (immature) death or selective abortion. From an ethical point of view most problematic in this study was the use of fetal material. Initially there was a lack of clarity about questions like: What was the duration of the pregnancy? Do the researchers use spontaneously aborted fetuses only? What tissues do they use, precisely? Will the fetal tissue be destroyed after being examined?

The research material according to this protocol is not anonymous. Thus, participants in this study, or their representatives, must be informed about the fact that the cell lines derived from their blood will be stored in order to retain the possibility of carrying out diagnostic research at a later time. During the IRB meeting the following question was raised: What precisely will be destroyed, if the patient changes her mind and decides to recall her consent: only her personal data or the cell line derived from her blood as well? In this case the patient was considered to have the right to ask that both her personal data and the cell lines derived from her blood be destroyed.

Case D: Cervical smear collection and analysis

The principal aim of this research proposal is to improve the detection rate of cytologic abnormalities in cervical smear preparations. The specific aims of this study are (1) to test the sensitivity of a new method of collecting and preserving cervical cells scraped from the uterine cervix for detection of cellular atypia and/or inflammatory conditions, and (2) to compare the overall adequacy of specimens collected by the new versus the routine procedures (Papanicolaou test). The study material consists of two parts. The cervix smears to be analysed for testing the new method are, for the most part, already in the possession of the researchers as a result of an earlier research project; these are considered the *normal* smears. However, the researchers also need some freshly prepared *pathological* smears. For this purpose they planned to make a cervix smear in women with pathology of the cervix, who present to the clinic for cytologic screening, as is usually done prior to a biopsy. They wanted to use the cell material that remained on the spatula device or brush after making this routine smear.

Again, in this study, the researchers initially thought that approval by an IRB was not necessary. They assumed that both parts of this research project did not involve human subjects, but only body material which is already available, which is normally not used and which otherwise would be discarded. An additional explanation for the researchers' misunderstanding was that for this study it was not necessary to know the patients' identity. The researchers wrote: "Since additional sampling is not requested from the subjects and only cells which are usually discarded are to be studied, a written consent form is not necessary".

With regard to the first part of this study, i.e., the renewed examination of older, normal smears, the following questions have been asked. Do patients know that their smears will be stored (and do they have to know it)? Do patients know that their smears will be used for this particular research objective (and do they have to know it)? Are the patients' smears used for other purposes than for which they initially were taken from their body? The answer to this last question is obviously 'yes': the smears routinely taken serve specific diagnostic purposes, whereas the same material in this particular research project will be used for a refinement of diagnostic automatized techniques.

Finally, retrospectively obtaining informed consent in the first part of the study was not regarded as feasible: how to trace the many women from which a smear has been stored? Moreover, it was argued that seeking their informed consent retrospectively could cause much needless commotion. On the other hand, it was unclear whether there had been any discussion with the patients regarding the fact that bodily material would be stored and used for future research purposes.

With regard to the second part of this study, similar questions may be asked. However, in practice, there is no obstacle to obtaining informed consent when making fresh smears. From an ethical and legal point of view this is regarded as an absolute condition for implementing the research protocol.

V. RESEARCH ETHICS AND HUMAN BODY PARTS

As may be concluded from the examples above, many problems arise in connection with biomedical research with human body parts: the safety of body products, the subject's privacy, the storage and use of body material (blood, tissue, and sperm banks), the difference between therapeutic and non-therapeutic experiments, the commercialization and exploitation of body parts and body products.

In several nations (e.g., the United States) the sale of biological materials in other contexts than transplantation medicine is allowed for the purposes of both research and therapy. Plasma providers are routinely reimbursed for the donation of this vital substance. Similarly, those who provide sperm for the purposes of artificial insemination are reimbursed for their donation [4]. In other countries, the general ethos is to disallow the sale of body parts such as blood or kidneys. This negative attitude towards commercial trade with bodily material may have been the underlying motive for the IRB to not approve the research protocol in case A. For example, the Health Council committee in the Netherlands (mentioned earlier) very strongly emphasised the importance of the principle of non-commercialism ([12], pp. 12, 34, 46, 53-54, 70, 88). This principle is taken to be so self-evident; hardly any argumentation is presented to defend it. The main argument is presented in the following one-liner: "The great objection is the danger of abuse: autonomy means little where the choice is between dying of hunger and selling a kidney" ([12], p. 34). It is also noted that others regard commercial considerations as incompatible with the fundamentally altruistic nature of voluntary donation. In a market system altruistic motives to donate blood, organs, and tissues may be undermined. If human body parts ought not to be collected, stored, or used with a view of making a profit, the question remains, however, whether people who donate body parts for scientific research or other purposes should be reimbursed. It is generally accepted practice in the Netherlands that no material reward be given to donors of blood, sperm, organs, and tissues. It is, on the other hand, argued that reasonable payments to cover expenses (including the subject's time and effort) are acceptable, particularly in cases of supplying material to the pharmaceutical industry (which will, perhaps, make a fortune) ([12], p. 70). Payments to compensate for expenses are in fact quite usual in the context of medical experiments. The distinction between selling body parts with a commercial motive and providing the same parts with an altruistic motive may in practice disappear when compensation for donors and research subjects is no longer modest.

Another important issue concerns the use of body material that contains information that is possibly traceable to the particular person concerned. As shown earlier, the collection, analysis, use, and storage of bodily material is common practice in biomedical research. Especially in genetic research and genetic counseling it is essential that a large and diversified amount of genetic material remains at the disposal of researchers and counselors [10]. Cell lines of genetic material throughout the world are used in diagnosis and research. As is the case with cancer research, these cell lines are indispensable for clarifying the disease processes concerned. Most clinical genetics centers have a cell bank, which serves, generally speaking, two functions: genetic diagnostics and biomedical research. If material will be used for diagnostic purposes, it is required that it is stored *by name* or at least contains information that is traceable to particular persons. However, identifiability is not always necessary if, for instance, the genetic center only uses the cell material for biomedical research.

The use of genetic material as well as fetal tissues (as illustrated in case C) have led to moral questions in particular, because of the sensitive nature of these bodily materials. Stricter conditions have been proposed for their use. For example, the Dutch committee has proposed that express consent be obtained for research with fetal tissue originating from abortions, even when the material is given anonymously ([12], p. 83).

The project referred to in case D may be used to clarify some ethical problems concerning (non-)anonymity. Because this was an international project which will also be carried out in the United States, the researchers are supposed to follow the *U.S. Guidelines for the Protection of Health and Human Services* [17]. The regulations concerned stipulate that an applicant organization has the responsibility of safeguarding the rights and welfare of human subjects. The regulations define 'human subject' as 'a living individual about whom an investigator conducting research obtains (1) data through intervention or interaction with the individual or (2) identifiable private information' [17]. The regulations extend to the use of human organs, tissues, and body fluids from identifiable human subjects as well as to graphic, written, or recorded information derived from individually identifiable human subjects. Exempt from coverage by these regulations is (among others) the following category: Research involving the collection or study of existing data, documents, records, pathological specimens, or diagnostic specimens, if these sources are publicly available or if the information is recorded by the investigator in such a manner that subjects cannot be identified, directly or through identifiers linked to the subjects [17]. As to these regulations a crucial question is, what 'publicly available' means. Furthermore, if individual subjects can not be identified via their body material, does this mean, that a researcher is allowed to use this material freely?

The Dutch Health Council committee formulates as one of the cornerstones of good practice the guiding principle, that "material used for purposes other than that for which it was taken should if possible not be traceable to the individual concerned" ([12], p. 53). Many uses of bodily material, at least more than in current practice, do not require identifiable material. If the identity of the donor or patient is relevant to the research goal, the committee proposes that express consent be obtained for any further use of the human tissue ([12], p. 71).

The discussion of the four cases illustrates that most ethical problems that arise in biomedical research with human body parts concern the legal and moral status of body parts. Generally speaking, it is the ownership of the body and body parts that is at stake. Contemporary medicine has focused on the human body as of instrumental value, a resource potentially beneficial to others: patients, physicians, researchers, commercial entrepreneurs. In his exploration of the property paradigm of the body, Campbell argues: "The image of the *body* as property relies on a sense that *parts* of the body, such as organs, gametes, or cellular tissues, can be transferred to, acquired by, and manipulated by others" ([3], p. 36). In this new context it is morally crucial how the pursuit of instrumental aspects of the body can be reconciled with the anthropological experience of human life as embodied existence, and of human persons as embodied selves, but also how the desired bodily benefits can be obtained without violating the body's intrinsic value.

With respect to the acquisition of body parts it is useful to distinguish two situations:

(1) Body material that is already available, because the patient has donated blood or undergone a biopsy in an earlier stage of diagnosis or in a previous research study. The patient's serum, a small part of his kidney, a smear of her cervical uterus, or a cell line derived from his cells are stored in a laboratory or tissue bank. In this situation the person who is storing or using the bodily material of other persons has to act in accordance with the original purpose for which that material has been

discharged or donated. The use of human body parts for other purposes than for which they were taken and for which consent had been acquired is not allowed. This general rule holds for the use of personal data, too. As case B illustrates, this rule also applies to so-called 'waste material'. Although the placentas used in this study represent anonymous material, the women concerned should be given general information and the opportunity to object to this particular use of their body parts.

(2) The body material is not yet available in the sense that it is still an integrated part of the body as a whole. To dispose of body material in this situation, it is necessary to carry out certain actions on the human body, i.e., the living body or even the dead body. Examples are the use of cadaveric brain tissue for research purposes or the cultivation of human cells of foreskins removed in circumcision for the production of skin to be used in reconstructive surgery. In the Netherlands, article 11 of the Constitution that guarantees the inviolability of the human body, applies to these cases: an explicit informed consent from the persons involved is required [8,9].

With respect to the ownership of human body parts, the question is how far a person's ownership and the sovereignty over his body parts actually reaches. From a legal perspective the question arises whether human body parts discharged from the body are to be seen as 'things' in the sense of the Law of Property (see [13]). Generally speaking, the answer is positive, albeit that some authors want to make an exception for special body parts, e.g., those body parts which do not regenerate after removal [8]. From an ethical point of view it is argued that people have the right to treat their own body parts as property, particularly their regenerative parts: people's body parts are their personal property [1]. This individual power to determine what happens with materials taken from the body also extends to the realm of waste materials such as sweat, urine, or discardable tissues. Patients or research subjects should not be anxious that once in hospital their bodies or body parts will be used instrumentally and without their expressed informed consent for the purpose of research or for commercial ends. At the same time they should be fully informed as to what will be done with any and all of their biological materials [4]. Regrettably, the practice of providing general information about the storage and use of human body parts is virtually non-existent in most research/health care institutions.

VI. CONCLUSION

Although biomedical research with human body parts is common in present-day medicine, it has received less discussion in the bioethical literature than medical experimentation with human beings. This is understandable because a person has a higher value than any of his or her body parts. Nevertheless, in the field of human research with human body parts a number of ethical problems arise and are in need of clarification. It is precisely due to the fact that human body parts 'belong to' or 'have belonged to' a person, that ethical questions about their proper use, collection, and storage are significant. Ultimately the ownership of the human body is at stake in research ethics.

Department of Ethics, Philosophy and History of Medicine, and
Center for Ethics
Catholic University of Nijmegen
Nijmegen, The Netherlands

NOTE

Translated from the Dutch: 'medisch-wetenschappelijk onderzoek waarvan deel uitmaakt het onderwerpen van personen aan handelingen of het opleggen aan personen van een bepaalde gedragswijze' ([20], art. 1b).

BIBLIOGRAPHY

1. Andrews, L.B.: 1986, 'My Body, My Property', *Hastings Center Report* **16** (5), 28-38.
2. Broxmeyer, H.E. *et al.*: 1992, 'Human Umbilical Cord Blood as a Source of Transplantable Hematopoietic Stem and Progenitor Cell', *Current Topics in Microbiology and Immunology* **177**, 195-204.
3. Campbell, C.S.: 1992, 'Body, Self, and the Property Paradigm', *Hastings Center Report* **22**, 34-42.
4. Caplan, A.L.: 1985, 'Blood, Sweat, Tears and Profits: The Ethics of the Sale and Use of Patient Derived Materials in Biomedicine', *Clinical Research* **33**, 448-451.
5. Capron, A.M.: 1991, 'Protection of Research Subjects: Do Special Rules apply in Epidemiology?', *Journal of Clinical Epidemiology* **44**, 81S-89S.
6. Corbitt, G., Bailey, A.S., Williams, G.: 1990, 'HIV Infection in Manchester, 1959', *Lancet* **336**, 51.
7. Dworkin, G. and Kennedy, I.: 1993, 'Human Tissue: Rights and the Body and its Parts', *Medical Law Review* **1**, 291-319.

8. Gevers, J.K.M.: 1989, 'Het gebruik van afgenomen lichaamsmateriaal in epidemiologisch onderzoek', *Nederlands Tijdschrift voor Geneeskunde* **133**, 173-75.
9. Gevers, J.K.M.: 1990, *Beschikken over cellen en weefsels,* Kluwer, Deventer, the Netherlands.
10. Gezondheidsraad: 1989, *Erfelijkheid: maatschappij en wetenschap. Over de mogelijkheden en grenzen van erfelijkheidsdiagnostiek en gentherapie,* Staatsdrukkerij en Uitgeverij, Den Haag, the Netherlands.
11. Harré R.: 1991, *Physical Being. A Theory for a Corporeal Psychology.* Blackwell, Oxford, UK.
12. Health Council of the Netherlands, Committee on Human Tissue for Special Purposes: 1994, *Proper Use of Human Tissue*, Health Council of the Netherlands (publication no. 1994/01E), The Hague, the Netherlands.
13. Heubel, F.: 1998, 'Defining the Functional Body and its Parts: A review of German law', in this volume, pp. 27-37.
14. James, J.: 1988, 'De cel in het organisme; over leven en dood', *Nederlands Tijdschrift voor Geneeskunde* **132**, 2348-2351.
15. Katz, J.: 1972, *Experimentation with Human Beings,* Russell Sage Foundation, NY.
16. Munzer, S.R.: 1994, 'An Uneasy Case against Property Rights in Body Parts', *Social Philosophy & Policy* **11** (2), 259-286.
17. *Protection of Human Subjects*: March 8, 1983, 45 Code of Federal Regulations, 46, Office of Protection from Research Risks Reports.
18. Rothman, D.J.: 1991, *Strangers at the Bedside. A History of How Law and Bioethics Transformed Medical Decision Making*, Basic Books, NY.
19. Spicker, S.F. *et al.*: 1988, *The Use of Human Beings in Research. With special reference to clinical trials*, Kluwer Academic Publishers, Dordrecht, the Netherlands.
20. *Gewijzigd voorstel van Wet inzake medisch wetenschappelijk onderzoek met mensen.* Tweede Kamer der Staten-Generaal, vergaderjaar: 1995-1996.

PART II

THE HISTORY AND CONCEPT OF BODY OWNERSHIP

DIEGO GRACIA

OWNERSHIP OF THE HUMAN BODY: SOME HISTORICAL REMARKS

I. INTRODUCTION

Am I the owner of my body? In what sense? Is it mine or is it me? What is the difference between proper and property, own and ownership? Am I my proper body, or have I the property of my body? Are there property rights in relation with the body? Can I donate or gift my own organs? Can I sell them? Should the state be permitted to expropriate some organs of my body in view of the common good? These are some of the questions raised by the new biomedical technology, specially organ transplants and genetic engineering. Bioethicists must consider these questions, and propose serious, clear and practical solutions. What is in danger is so important that we must be extremely cautious. And one of the first things we should do in such a situation is to look back and see if we can learn from the lessons of history.

The question of the ownership of the human body is not new. Slavery and servility were accepted by many cultures throughout the world, and in Western culture; both were accepted until very recently. Slaves did not own their own bodies, nor did they possess property rights. Only with the liberal revolutions of the eighteenth century did the idea that all human beings are equals and deserve the same consideration and respect begin to be accepted by political institutions. "Every man has a property in his own person", wrote John Locke in the seventeenth century, when civil and political rights began to appear. But is this mastery or ownership complete? Can we dispose of our proper body as we do of other things? What are the differences between 'me', 'mine', and 'my property'?

In the first part of this essay I analyze some classical answers to the question of the ownership of the human body. In the second part, I analyze the nuances added by the 20th century to this issue. I propose an integrative model to approach the problem of the ownership of the human body.

H.A.M.J. ten Have and J.V.M. Welie (eds.), Ownership of the Human Body, 67–79.

II. WHO OWNS THE HUMAN BODY? – SOME CLASSICAL ANSWERS

In the history of Western culture there have been at least five different views concerning the question of the ownership of the human body; each of them with concrete and specific ethical consequences. The first and the second considered God as owner of the human body; the third and fourth believed that human being is the owner; and the fifth, society.

1. According to the first view, nature has an intrinsic order that is the norm of morality. This was the point of view of Greek philosophy. Nature has been ordered by God, and therefore is always right and beautiful. Only the unnatural can be considered wrong, bad, and ugly. Disease, for instance, is an unnatural disorder; the reason why the surgical amputation or mutilation of a limb is permitted. The preservation of the natural order, and its restoration when lost, are moral duties. This was the message of Socrates' death, and also of his moral philosophy, as explained in the Platonic dialogues. God alone is wise, and man can only be wise by imitating Him. This is the goal of the philosopher, the imitation of God. The way to know God is through the natural order, and therefore the philosopher must adopt his life to it. This ideal is specially clear in Stoicism, the philosophical movement that attempted to live in accordance with divine reason expressed in nature. This divine order, written in nature, is the supreme criterion of morality.[1] Moral virtue is the capacity to live in accordance with nature, with total indifference to pleasure and pain, health and disease, life and death. Man is not proprietary of his life, only administrator and warden, and must accept natural events magnamiously. Life is not a private property of individuals, but a gift of nature and God.

This naturalistic tradition never considered the legal concepts of ownership or property applicable to the human body, at least when dealing with free people. According to Roman Law, the human body was not something external, and therefore there was no sense in talking about 'dominion' or 'property', *quoniam dominus membrorum suorum nemo videtur* (Ulpian, *Dig* X, 2, 13). According to Ulpian, man is not the owner of his members, and therefore he is not the owner of his body. The possessor of the body is God, creator of life, and man is the *administrator et custos* of this gift. This in part explains the traditional legal and moral prohibition against self-mutilation and suicide, until quite recently. The living human body was considered in Roman Law as a constitutive element of each person, and not a 'thing'. Only the dead body was considered a 'thing', *res*, but *res religiosa*, and therefore *sui generis*, neither appropriable (*res extra patrimonium*) nor salable (*res extra commercium*), 'an object of cult and respect, rather than an object over which one may dominate: with the corpse, the only licit activity that can be considered is to provide its funeral honors and appropriate sepulture' ([12], p. 150). This, of course, means that the living body can not be an object of commerce. This is the reason why a sentence of the Digest reads: *liberum corpus nullam recipit aestimationem* (*Dig* 9,3,7, *in fine*) – the body of free people has no price.

Roman Law was strongly influenced by Stoicism, the philosophical school of thought that affirmed and defended the sacredness of each human life and human body. *Homo sacra res homini*, wrote the Stoic philosopher Seneca (*Ep* 95, 33): the human being is something sacred, and therefore man has no dominion over his body, only disposition. And this disposition must be carried out according to the Greek virtue of *hosiótes* (the Latin *pietas*). This also explains why the human being has a moral (and legal) duty to respect the body, and also why the dead body was considered 'inviolable' and 'unsalable'.

Roman Law distinguished clearly between 'possession' (*possessio*), and 'disposition' (*dispositio*). The first could be complete or 'dominion' (*plenum dominium*), and incomplete or 'property' (*nuda proprietas*) [10]. The owner had the right of property over a thing, but he was not its *dominus*, lord, or master. In a paternalistic society, like the Roman, only fathers (*domini*) had dominion over their progeny. When this patrimony could no longer be managed by a person or family, the lord, for instance, the Emperor would distribute the property between different persons or families, but always retaining dominion. Only God had complete dominion over things: He was the first Lord of things. The Emperor had also power of dominion, and therefore was considered the second Lord; the father (*pater familias*) of a private house (*domus*), was the third one. No one else could have dominion over things.

Different from the concept of 'possession' is 'disposition'. A person may dispose of things without being their owner, or without having dominion over them. We can distinguish three kinds of disposition, however: use (*usus*), enjoyment (*fruition*), and the combination (*usus fructus*). Generally, the possessor of a thing was the person who held dominion over it. But this was not always the case; sometimes the person who held dominion was different from the one who had disposition.

In short, Roman Law distinguished between dominion and property; only the first could be complete, or perfect. The use or disposition of things was not identical to dominion and property, and therefore was not considered absolute.

2. The three monotheistic religions, the Jewish, the Christian, and the Muslim, promptly accepted the Greek naturalistic point of view. This was the beginning of the three theologies, the application of the Greek approach to the Sacred books [18]. All of them consecrated naturalism, considering natural order from a supernatural perspective, as expression of God's will. This permits understanding of the multi-secular success of the naturalistic approach in these three cultures.

Thomas Aquinas exposes in *Summa Theologica* the question whether the mutilation of a limb of the human body can be licit.[2] His answer was that limbs are always part of a whole, and therefore mutilation can be justified only if it benefits the whole. As such (*per se*), mutilation is unjustifiable and immoral. Only circumstantially (*per accidens*), for instance, in case of a grave disease, when the diseased limb can endanger the good of the whole, amputation can be considered licit – "if the patient consents to the intervention" (*de voluntate eius cuius est membrum*), or if the surgeon believes that the intervention is medically appropriate. The same is true when the conflict arises between a member of society and the political or civil community. Here also the common good must prevail over the private and individual good, but in this case mutilation or death can only be authorized by the political power, not by private person. And Thomas Aquinas concludes his argumentation affirming that in all other cases mutilation is always illicit.[3]

This text has been of great historical importance, not only because it gave intellectual and moral justification to capital punishment for religious, moral, and political reasons, but also because it considered the mutilation of human beings permissible in two exceptional cases: for medical and for social or political reasons. A private person can not morally dispose of all or of part of his or her body; only in view of the common good, can mutilation be justified. In other words, man is not the owner of his body. The state has greater power over his body than he himself. This 'contradiction', completely logical from the point of view of naturalism, will become inverted by the liberal thinkers of the modern world.

3. The naturalistic approach was criticized in modern times, and specially in the Eighteenth century, when philosophers began to affirm that judgments about facts should always be of a posteriori synthetic character, and therefore uncertain. The empiric judgments are necessarily made *a posteriori*, the reason why they can only be true when all the possible cases have occurred, what in experience is impossible. Therefore, naturalism's belief in the capacity of knowing absolutely what nature 'is', and to deduce from it the norm of morality (what we 'ought' to do) is mistaken, and today is known as the 'naturalistic fallacy' ([17], pp. 469-70). Nature can not ground morality. Only man, the human being, is the true subject of rights and duties. John Locke, one of the fathers of the modern liberal thinking, says: 'Every man is his own lord'. Therefore, the body is the first and most important *property* of human beings, and also the means of acquiring all other things. The human being is not only the administrator of his own body, but also its proper master and owner. This permitted a restatement of the classical doctrine of Pandect, in the sense that the human being is *dominus* or lord of his or her body. Man is 'over' and not 'under' his body; in this way he becomes his 'owner'. In this view, the body becomes the primary property and the origin and ground of all other rights of property (self-ownership) ([6,7,11]).

This is perfectly evident, for example, in Locke's *Two Treatises on Civil Government*. According to Locke, the first and principal title of property is work, means by which every human being 'embodies' or 'incorporates' and makes the things in which he puts all his effort. The primary and principal property is the body, and the secondary, all those things incorporated by means of work. These are his words:

> Though the earth, and all the inferior creatures, be common to all men, yet every man has a property in his own person: this nobody has any right to but himself. The labor of his body, and the work of his hands, we may say, are properly his. Whatsoever then he removes out of the state that nature hath provided, and left it in, he hath mixed his labor with, and joined to it something that is his own, and thereby makes it his property ([24], Book II, c.V, *On Property*, §27).

Therefore, the human body is 'private property' by excellence, and everything 'embodied' by his effort and work enters inside the bounds of the rights of 'property' and 'privacy'. The human body is 'inviolable' by the state and by society, but not by the owner-person. He or she is the owner of his or her own body, and therefore he or she can dispose of it according to his or her own will, e.g., taking his or her own life away. The paradigmatic example of this was the essay by David Hume, *Of Suicide*. Hume tries to prove that suicide does not have a 'criminal nature', and that the laws that consider it a crime are wrong [16]. The human body is not inviolable; therefore, the person who disposes of it consciously and voluntarily is not committing a crime. And if it is violable, it seems we must conclude that it is also alienable or salable, at least in part. The human body as with any property has a price.

4. Discussing Hume, Kant began a different attitude, of great success in European philosophy of the nineteenth and the twentieth centuries. Kant considered reason capable of founding morality in a categorical and imperative form. This formal principle says, that "we should always act so as to treat Humanity, whether in our person or in that of another, as an end, and never as a means only" ([20], p. 91). The categorical imperative obliges absolute respect of all human beings, including our proper person. Ethics is concerned not only with the duties every human being has toward others, but also with the duties everyone has with himself. Both of them are pure duties, and therefore different from other kinds of individual goals, like happiness or self-interest. Kant says explicitly in the *Lectures on Ethics*, that "the principle of self-regarding duties is a very different one, which has no connection with our well-being or earthly happiness" ([21], p. 117). The duties towards ourselves are not founded in favor but in self-esteem. That means, Kant says, that our actions must be in keeping with the worth of man ([21], p. 124). This worth is proper of the person as a whole, and consequently includes his body. Our life is entirely conditioned by our body, so that we cannot conceive of a life not mediated by the body, and we cannot make use of our freedom except through the body ([21], pp. 147-148). The body is the means of realization of our freedom, viz., of our moral duties, and therefore we must respect it.

A human being, then, is not entitled to sell his limbs, even if he were offered ten thousand thalers for a single finger ([21], p. 124). If he were entitled, he could sell all his limbs. But this is not the case. We can dispose of things which have no freedom but not of a being which has free will ([21], p. 124).

Persons are not things. Things can be treated as means; persons are ends in themselves. A man who sells himself makes of himself a thing and, as he has jettisoned his person, it is open to everyone to deal with him as they please ([21], p. 124).

The consequence of this reasoning is that man has no property rights over his body. Kant explains this explicitly when dealing with the 'Duties towards the body in respect of sexual impulse'. He writes:

> Man cannot dispose over himself because he is not a thing; he is not his own property; to say that he is would be self-contradictory; for in so far as he is a person he is a Subject in whom the ownership of things can be vested, and if he were his own property, he would be a thing over which he could have ownership. But a person cannot be a property and so cannot be a thing which can be owned, for it is impossible to be a person and a thing, the proprietor and the property. Accordingly a man is not at his own disposal. He is not entitled to sell a limb, not even one of his own teeth ([21], p. 165).

The only right of disposal of bodily parts that the human being has, is when it is necessary for his own self-preservation. For instance, in order to restore health and to secure self-preservation, amputation of limbs is permitted.

> We may treat our body as we please, provided our motives are those of self-preservation. If, for instance, his foot is a hindrance to life, a man might have it amputated. To preserve his person he has the right of disposal over his body ([21], p. 149).

This doctrine about the body was maintained by Kant throughout his life. In his last ethical work, the second part of the *Metaphysics of Morals*, entitled 'The Doctrine of Virtue', he writes:

> To dispose of oneself as a mere means to some end of one's own liking is to degrade the humanity in one's person (*homo noumenon*), which, after all, was entrusted to man (*homo phenomenon*) to preserve ([19], p. 423).

And later on:

> To deprive oneself of an integral part or organ (to mutilate oneself) e.g., to *give away* or *sell* a tooth so that it can be planted in the jawbone of another person, or to submit oneself to castration in order to gain an easier livelihood as a singer, and so on, belongs to partial self-murder.

> But this is not the case with the amputation of a dead organ, or one on the verge of mortification and thus harmful to life. Also, it cannot be reckoned a crime against one's own person to cut off something which is, to be sure, a part, but not an organ of the body, e.g., the hair, although selling one's hair for gain is not entirely free from blame ([19], p. 423).

The body is the incarnation of reason, and therefore the place in which reason, the image of God, is seated. The body is not 'mine', it is 'me'. In other words, my body is not a thing, but a 'person' I can not dispose of. This is the Kantian doctrine. Its influence in the development of the ethical and legal analysis of western culture has been enormous. It helps us understand our reluctance to speak about ownership of human bodies ([15,5,22]).

5. Consider now the view that there is no strict distinction between 'property' and 'person'. From the word 'own' (personal) we get to 'owner' (proprietary), and to 'ownership' (property). One of Hegel's disciples, Max Stirner, in *Der Einzige und sein Eigentum* [26], defends, against socialism, the concept of 'property' (*Eigentum*) as ground of 'that proper to each one' (*der Einzige*), that is to say, of each one's personality. Personality is then understood as the capacity of being owner. Against Stirner, and more specifically challenging his idea of property as ground of personality, Marx and Engels produced *Die deutsche Ideologie*. They say that what Stirner defends is no different from the "most old and trivial bourgeois objections" ([25], p. 210), already exposed by the father of the French ideological movement, Destutt de Tracy, who, in his *Traité de la volonté* [9], had made the human body the point of encounter of the *proprieté* (private propriety) and the *proper* (personality), "what Stirner makes through the word riddle of *Mein* and *Meinung*, *Eigentum* and *Eigenheit*" ([25], p. 211). What Destutt de Tracy had tried to demonstrate, and what Max Stirner had accepted, was "that *proprieté*, *individualité* and *personnalité* are identical things and that in the *moi*, the *mien* also is implicit" ([25], p. 210). For Marx and Engels the bourgeois always tend to identify personality with private property, and think that

> when destroying the property, that is to say, when destroying my existence as capitalist, as landowner, or as producer, and your existence as workers, you destroy my individuality and also yours; when you restrict me from exploiting you as workers, to inhibit

> reimbursing my profits, my credits, or my rents, you incapacitate me as an individual; therefore, when the bourgeois tells the communist: when abolishing my existence *as bourgeois*, you destroy my existence *as individual*, when he identifies himself that way, as bourgeois, with himself as individual, we have to recognize his sincerity and shamelessness. About the bourgeois, it really happens that way: he only believes being a real individual is being a bourgeois. But in the moment in which the bourgeois theorists give this statement a general meaning, identifying also theoretically the bourgeois property with individuality (trying to justify logically this identification), foolishness starts acquiring a solemn and sacred tone ([25], p. 211).

I have transcribed this long paragraph because it shows clearly the final aim these analyses seek: the discovery of the concept of *Entfremdung*, or alienation. Marx thought that the binomial was not *Eigene-Eigentum*, as Max Stirner said, but *Fremde-Entfremdung*. The private property of the means of production, for Marx, not only does not personalize, but it depersonalizes: it turns the one's own into alien (*das Eigene als Fremdes*). Against this, as Alfred Kurella says in his book, *Das Eigene und das Fremde* [23], socialism tries to do exactly the opposite, turn the alien into one's own (*das Fremde als Eigenes*). This can only be achieved by socializing the property of the means of production, in which we stress human work, the human body. The body has the peculiarity of being at the same time a good of consumption and a good of production. Because of the first quality, it has 'private' and 'individual' character; due to its second, it is 'public' and 'social'. According to Louis Blanc's famous phrase, quoted by Marx in *Critique of the Gotha program*, 'from each according to his capacity; to each according to his necessities', we could state that the public dimension of the body is its 'capacities', and the private one its 'necessities'. And as everything related with health belongs to the first of these fields, we have to conclude that the human body is not only private property, but society's property. If for the liberal theory the'individual body' was the means of appropriation and personalization of the 'social body', here it is exactly opposite: the 'social body' is the ground for the appropriation and personalization of the 'individual body'. Understanding the human body as 'private property' and defending it by appeal to the right to privacy is the main source of 'alienation' or 'depersonalization' of the human being. The personalization of the human body starts with the acknowledgement of its essential public and social dimension. Instead of the absolutization of the right to privacy defended by the liberal movement, the human body must have first a public and social dimension. The ethics of the human body is not for Marx, primarily individual but social, and many of the questions that worry us today, such as the ownership of the organs used in transplants or the ownership of the human genome, must be handled from a social instead of an individual point of view. According to this position, the human body is violable.

We can therefore see that there have been at least five historical answers to the question concerning the ownership of the human body: the two classical (Greek and Christian), agree in that God is the 'owner' of the body, and consider it inviolable and inalienable; the liberal answer states that the owner is the specific individual, for whom his body is violable and alienable; the Kantian establishes that persons are the real owners of their bodies, but considers that this ownership does not permit its free violation and alienation; finally, the socialist approach holds that the principal owner is society, for whom the human body is considered violable.

III. THE 20th CENTURY: TOWARDS AN INTEGRATIVE ANSWER TO THE QUESTION OF THE OWNERSHIP OF THE HUMAN BODY

The way in which in western Europe, and today, at the end of the twentieth century, we understand the question of the ownership of the human body is not independent of the classical answers exposed above, but at the same time it is different from all of them. It is a new attitude though it had its origin in them, 'integrating' their results. From the Kantian position we learned that personal being is inviolable. Some will see in this personal characteristic of the human being the image of God while others will not, but many will accept it as the ground on which their morality stands, what makes us owners of rights and bearers of duties. These rights and duties have, at the same time, two different dimensions: an individual one, perfectly understood by liberal thought, and a social one, emphasized by the socialist tradition. Today we know that none of them is absolute, and that our moral life is in danger when it is reduced either to the exclusively private or the social. Moral life requires keeping both of them always in mind. From a certain point of view they seem complementary, from another they are clearly opposite; this also helps us understand why moral conflicts are increasing in our society. Mankind has never before adopted such a comprehensive moral model, and never before have moral controversies been so ubiquitous. The issue here, the ownership of human body, is a paradigmatic illustration of this postmodern state of affairs.

The primary role of the state is not to promote values, but to make sure that people have the opportunity to carry out their life projects. This is why its primary obligation is to avoid maleficence, and even injustice. This is the main aim of public law. When the state does not fulfill this task, there is no doubt it is allowing and promoting the expropriation and dispossession of the health and the body of its citizens. For example, a state that does not protect adequately the physical integrity of its members, or their equal opportunities of access to health services, will be an expropriator and dispossessor of the body of its citizens. Here the role of the state must be positive, avoiding maleficent and unjust behaviors. On the other hand, its role in the second level must be merely negative, allowing its members to carry out their life projects and to seek perfection and happiness. Whenever the state acts more deeply at this level, manipulating citizens' systems of values, the state is acting as dispossessor and expropriator.

As we have defined health as the process of possession or appropriation of one's own body, the levels we have established can be directly applied to the question we are here discussing: the ownership of the human body. The body is always 'my' body, it identifies with me, with my personal identity, with my I, and it is the instrument with which I can carry out my life projects, defined by principles like autonomy. But at the same time, that body has besides the individual and non-transferable dimension a generic one: it belongs to the human species, and it becomes completely useless and senseless when withdrawn from the context of human beings. Opposite the individual dimension of the body is also its specific dimension. My body is one in the great 'set' of human bodies, different from them all, but at the same time inseparable from the rest. This is the reason why I have certain moral obligations toward them all, perfectly stated by the principle of non-maleficence. I cannot act maleficently on the bodies of others, nor can they on mine; I must try to extend this non-maleficence to all the members of the human species.

Consequently, besides the individual ethical obligation I have to own my body autonomously, in accordance with my ideals of perfection and happiness, I also have a specific ethical duty not to act maleficently toward the bodies of others. These rules and prohibitions must be the same for everyone. Even though our body is ours, we all have a duty not to act maleficently toward the bodies of others, and the state must ensure that each of us is protected.

Complutense University
Madrid, Spain

NOTES

[1] Cicero, *De legibus* I,6: "Lex est ratio summa, insita in natura, quae iubet ea quae facienda sunt, prohibetque contraria".

[2] "Utrum mutilare aliquem membro in aliquo casu possit esse licitum" ([27] q. 65, a.1).

[3] "Cum membrum aliquod sit pars totius humani corporis, est propter totum, sicut imperfectum propter perfectum. Unde disponendum est de membro humani corporis secundum quod expedit toti. Membrum autem humani corporis per se quidem utile est ad bonum totius corporis: per accidens tamen potest contingere quod sit nocivum, puta cum membrum putridum est totius corporis corruptivum. Si ergo membrum sanum fuerit et in sua naturali dispositione consistens, non potest praecidi absque totius hominis detrimento. Sed quia ipse totus homo ordinatur ut ad finem ad totam communitatem cuius est pars, ut supra dictum est, potest contingere quod abscisio membri, etsi vergat in detrimentum totius corporis, ordinatur tamen ad bonum communitatis, inquantum alicui infertur in poenam ad cohibitionem peccatorum. Et ideo sicut per publicam potestatem aliquis licite privatur totaliter vita propter aliquas maiores culpas, ita etiam privatur membro propter aliquas culpas minores. Hoc autem non est licitum alicui privatae personae, etiam volente illo cuius est membrum: quia per hoc fit iniuria communitati, cuius est ipse homo et omnes partes eius.
Si vero membrum propter putredinem sit totius corporis corruptivum, tunc licitum est, de voluntate eius cuius est membrum, putridum membrum praescindere propter salutem totius corporis: quia unicuique commissa est cura propriae salutis. Et eadem ratio est si fiat voluntate eius ad quem pertinet curare de salute eius qui habet membrum corruptum. Aliter autem aliquem membro mutilare est omnino illicitum" ([27] q. 65, a. 1).

BIBLIOGRAPHY

1. Boorse, C.: 1975, 'On the distinction between disease and illness', *Philosophy and Public Affairs* **5**, 49-68.
2. Boorse, C.: 1976, 'What a theory of mental health should be', *Journal of the Theory of Social Behavior* **6**, 61-84.
3. Boorse, C.: 1976, 'Wright on functions', *Philosophical Review* **85**, 70-86.
4. Boorse, C.: 1977, 'Health as a theoretical concept', *Philosophy of Science* **44**, 542-573.
5. Chadwick, R.F.: 1989, 'The Market for Bodily Parts: Kant and Duties to Oneself', *Journal of Applied Philosophy* **6**(2), 129-139.

6. Cohen, G.A.: 1986, 'Self-Ownership, World-Ownership, and Equality', in Frank S. Lucash (ed.), *Justice and Equality Here and Now*. Cornell University Press, New York and London, UK.
7. Cohen, G.A.: 1986, 'Self-Ownership, World-Ownership, and Equality: Part II', *Social Philosophy and Policy* **3**, 77-96.
8. Daniels, N.: 1985, *Just Health Care*. Cambridge University Press, Cambridge, MA.
9. Destutt de Tracy, A.L.C.: 1826, *Traité de la volonté et de ses effets*. Paris.
10. D'Ors, A.: 1992, *Elementos de Derecho Privado romano* (3rd ed.). Eunsa, Pamplona.
11. Fowler, M.: 1980, 'Self-Ownership, Mutual Aid, and Mutual Respect: Some Counter-examples to Nozick's Libertarianism', *Social Theory and Practice* **6**, 227-245.
12. Gordillo Cañas, A.: 1987, *Transplante de órganos: 'pietas' familiar y solidaridad humana*. Civitas, Madrid.
13. Gracia, D.: 1991, *Introducción a la Bioética*. El Buho, Bogotá.
14. Gracia, D.: 1991, *Procedimientos de decisión en ética clínica*. Eudema, Madrid.
15. Harré, R.: 1987, 'Body obligations', *Cogito* **1**, 15-19.
16. Hume, D.: 1964, 'Of Suicide', in Thomas Hill Green and Thomas Hodge (eds.), *The Philosophical Works of David Hume*. Scientia Verlag, Aalen. (Reprint from the edition of London 1882), Vol. IV, pp. 406-414.
17. Hume, D.: 1888, *A Treatise of Human Nature: Being an Attempt to Introduce the Experimental Method of Reasoning into Moral Subjects*. Clarendon Press, Oxford (reprint 1967).
18. Jaeger, W.: 1961, *Early Christianity and Greek Paideia*. Harvard University Press, Cambridge, MA.
19. Kant, I.: 1907, *Die Metaphysik der Sitten*, in *Kant's gesammelte Schriften*, Vol. VI. Georg Reimer, Berlin.
20. Kant, I.: 1948, *The Moral Law* (English translation of the *Groundwork of the Metaphysic of Morals* of I. Kant, by H.J. Paton). Hutchinson, London, UK.
21. Kant, I.: 1963, *Lectures on Ethics*. Harper & Row, New York.
22. Kass, L.R.: 1992, 'Organs for Sale? Propriety, Property, and the Price of Progress', *The Public Interest* **107**, 65-86.
23. Kurella, A.: 1970, *Das Eigene und das Fremde: Neue Beiträge zum sozialistischen Humanismus*. Aufbau Verlag, Berlin / Weimar, Germany.
24. Locke, J.: 1823, *Two Treatises of Government*, in *The Works of John Locke*. Scientia Verlag, Aalen, (reprinted 1823), Vol. V, pp. 353-354.
25. Marx, K. and Engels, F.: 1969, *Die Deutsche Ideologie*, In *Marx-Engels Werke* (Vol. III). Dietz, Berlin.
26. Stirner, M.: 1845, *Der Einzige und sein Eigentum*. Kaspar Schmidt, Leipzig.
27. Thomas Aquinas: 1963, *Summa Theologica*, Vol. III, *Secunda secundae* (3rd ed.). Biblioteca de Autores Cristianos, Madrid.

ZBIGNIEW SZAWARSKI

THE STICK, THE EYE, AND OWNERSHIP OF THE BODY

I. INTRODUCTION

It is a tradition in some countries that blind people use a white stick. The white stick is not only a useful instrument for spatial orientation but it has also symbolic meaning. If you see a white stick you know that there is somebody who is physically handicapped. Someone who robbed a blind man of his stick would certainly be viewed as a moral monster. We would condemn his act not only because the stick was the blind man's property or because it would expose him to unnecessary suffering. Our reaction to stealing the stick from a blind person is so strong because we perceive the stick as an integral part of the blind person and his world. The stick defines him and his situation in the world. Robbing the blind man of his stick forces a partial disintegration of his self and the world around him.

Suppose now that we approach the blind man and offer to buy his stick for a generous sum. If he fully understands the transaction and without the slightest coercion consents to it, then it seems that there are no relevant moral reasons to disapprove of the transaction. Though it would be morally repugnant to steal the stick, there seems to be nothing morally wrong in buying it (his consent undercuts the moral problem). Why is it, then, that the very idea of buying an eye from a healthy person is so abhorrent to most of us? Why, if we accept trading in white sticks, have we such strong reservations against legalizing the market for human organs? What, if anything, is the moral difference between buying the living human eye and buying the blind man's stick?

II. THE STICK

It is not difficult to explain why we do not see anything wrong with the idea of selling the white stick.

The stick (like an eye removed from a person) is *a thing*. It can be brown or white, long or short, heavy or light. It can be seen and touched.

H.A.M.J. ten Have and J.V.M. Welie (eds.), Ownership of the Human Body, 81–96.

Whatever are its qualities it exists outside of the subject in objective, public, and unitary space and hence it is easily accessible for others. It belongs to the common world of material things that we can experience and use, and therefore it can be the object of different acts and actions of various individuals.

But if it is the blind man's stick it means that it is something more than a mere thing, a piece of branch cut off a tree. It is *an instrument* serving the realization of a particular purpose. As all instruments it mediates between the person and the external world. If the human body is a sum of potentialities, the function of the instrument is to develop and extend these potentialities. The stick is then an extension of the human hand. If I have in my hand a solid stick I feel more secure in a deserted area than with my bare hands alone. In the case of the blind man the stick, by extending the length of his hand, replaces his sight. It is for him something much more important than an umbrella when it is raining. If he is blind the stick defines the frontiers of his 'visual field'. He 'sees' the world with and through his stick and he remains in a special relation to his stick.

> "The blind man's stick" – says Merleau-Ponty – "has ceased to be an object for him, and is no longer perceived for itself; its point has become an area of sensitivity, extending the scope and active radius of touch, and providing a parallel to sight. In the exploration of things, the length of the stick does not enter expressly as a middle term; the blind man is rather aware of it through the position of objects than of the position of objects through it. The position of things is immediately given through extent of the reach which carries him to it, which comprises besides the arm's own reach the stick's range of action. If I want to get used to a stick, I try it by touching a few things with it, and eventually I have it 'well in hand', I can see what things are 'within reach' or out of reach of my stick. ... To get used to a hat, a car or a stick is to be transplanted into them, or conversely, to incorporate them into the bulk of our own body" ([18], p. 143).

But however close is the link between the person and the instrument there is always *a boundary* between her body and the instrument. The white stick as an instrument belongs to the external world of things, and although it fulfills a basic function in extension of the field of sensual perception of the blind man it is not and never will be a part of his phenomenological, lived body.

As a product of human labor the stick is always someone's actual, former, or future *property*. If the blind man owns the stick he may do with it what he wishes, other things being equal. He may sell, donate, or even destroy it. The clause 'other things being equal' is in this case exceptionally important because his right to property may be, in some situations, drastically restrained. If the stick the blind man owns was, e.g., once property of Homer it may indeed be an irreparable loss to destroy it. Similarly, if the blind man used his stick to attack savagely a child, we would have no hesitation in disarming him. He has simply no right of making harmful use of his property. By imposing the appropriate laws the state may significantly restrict our right to use our property.

I said in the introduction that the stick defines the blind man and his situation in the world. Actually that is not quite true. What is really essential in his predicament is his blindness. If by selling his stick he could have bought a pair of post mortem recovered cornea to regain his sight by a successful operation, probably nobody would condemn such a transaction. Certainly he would be better off getting his sight back and nobody would deny our vital 'welfare interest'[9] in preserving or recovering our sight. What is wrong, then, if in the situation of substantial shortage of post mortem transplants he is trying to buy a cornea from a living donor? If we accept donation or trading in post mortem recovered tissues and organs why do we have objections against selling a healthy human eye? What is then so special about the status of the living eye?

III. THE EYE

The living human eye is not primarily a thing. Of course, I can touch my eye or see it in the mirror and in this sense it *appears* to me, like the eyes of other people, as a thing, a part of the body, a solid organ of visual perception, or another object of the external world. But this is only a secondary way of perceiving my eye. By its very nature my eye cannot exist as a living eye outside of me, outside of my body, and I cannot experience it the way I perceive other objects of the external world.

> "I am the *Other* in relation to my eye" – writes Sartre. "I apprehend it as a sense organ constituted in the world in a particular way, but I cannot 'see the seeing'; that is, I cannot apprehend the process of revealing an aspect of the world to me. Either it is a thing among other

> things, or else it is that by which things are revealed to me. But it cannot be both at the same time" ([22], p. 219).

If I exist as a person I exist due to the body which is equipped with a sensual system that orientates us in the world. Only man, due to his upright posture and a specific location of the eyes is able to behold the world. "The animal is confined to the limits of its own body" ([25], p. 340); the animal sees the world but only people can admire it. If we were, as a species, deprived by nature of sight our 'picture' of the world would be drastically different, argues Hans Jonas [13] in *The Nobility of Sight*. We would be devoid of the most fundamental categories of our conceptual framework. We would miss the concept of shape, picture, image, horizon, spacial distance and, what is the most important, we would lack some basic philosophical concepts.

> "*Simultaneity of presentation*" – says Jonas in his conclusions – "furnishes the idea of enduring present, the contrast between change and the unchanging, between time and eternity. *Dynamic neutralization* furnishes form as distinct from matter, essence as distinct from existence, and the difference of theory and practice. *Distance* furnishes the idea of infinity. Thus the mind has gone where vision pointed" ([13], p. 328).

That sight is indeed the most important and the noblest of all senses and the living eye, that most essential attribute of our individual and species life, is definitely one of the most relevant physical dimensions of our personhood. And perhaps this makes the living eye (next to the human hand) so special in the history of humanity.

It is also doubtful that we may regard the living eye as an instrument which can be used the same way ordinary tools are applied. Though the eye is indeed the organ of sight (and the Greek etymology of the word [*organon* = tool] suggests its instrumental character) it is misleading to treat the eye as an instrument. All instruments have their users. If the eye were an instrument we would have to decide who is its user. But, as it has been convincingly shown by phenomenological analysis [12,17,22] it is not the case that our bodies or sensory organs are specific instruments of the soul or mind. If the body was an instrument it would have to be used by another more primordial body and this implies a regress *ad infinitum* ([17], p. 209). Only if we distinguish between two modi of existence of the body – the body-for-itself and the body-for-others – is it possible to talk about the instrumental use of the Other's body. Sartre makes a fine point when he describes a situation in which "I cause the nails to be pounded in by the Other that the hand and the arm become in turn instruments which I utilize and which I surpass toward their potentiality" ([22], p. 229). But it is the hand of the Other which is the instrument, not my hand.

> I do not apprehend *my* hand in the act of writing but only the pen which is writing; this means that I use my pen in order to form letters but not *my hand* in order to hold the pen. I am not in relation to my hand in the same utilizing attitude as I am in relation to the pen; I *am* my hand ([22], p. 229).

The same, in the case of my eyes. When I study a painting or watch a movie on TV with my gaze fixed on the crashing planes I am not using my eyes in order to see the picture or to watch TV – *I am all in my eyes*, and my body as well as my other senses are simply reduced to a "background attitude" ([14], p. 24).

My living eye is thus not a thing, an instrument, nor a property, either. Property is both a moral relation between persons and a social convention and institution. Property appears where there is a human community in which individuals produce, exchange, inherit or take possession of things. Property is a certain moral institution which is defined by constitutive rules [11]. I cannot say that something is my property unless I presume the existence of a certain set of rules determining my behavior as well as the behavior of others towards that thing. Frank Snare [23], who follows the general line of Rawls, put forward several necessary conditions for the definition of property. *A owes P if and only if he has a right to use P, if this is the exclusive right of A and if he can transfer this right to other parties.* There seems to be nothing new in this analysis of property but what is interesting is the way Snare defines these three rights:

1. *Right of Use:*
   A has a right to use P, i.e.,
   (a) It is not wrong for A to use P, and
   (b) It is wrong to others to interfere with A's using P.
2. *Right of Exclusion:*
   Others may use P if and only if A consents, i.e.,
   (a) If A consents, it is *prima facie* not wrong for others to use P;
   (b) If A doesn't consent, it is *prima facie* wrong [for others] to use P.
3. *Right of Transfer:*

A may permanently transfer the rights in Rules 1 and 2 to specific other persons by consent.

Now if we define property in terms of a right to use, can one argue that because I can use myself I am the owner of my body? Locke had no doubt that it this the case. In a celebrated passage he wrote: "Though the earth and all inferior creatures be common to all men, yet every man has a property in his own person; this nobody has any right to but himself. The labor of his body and the work of his hands we may say are properly his" [16]. But Locke was wrong; for at least two reasons:

First, because, as it was pointed out by J.P. Day [7], he confused two meanings of the pronoun *my*. It is possible to distinguish between possessive and non-possessive use of *my*. When I say this is my house, this is my pen or my paper, I use *my* as the possessive pronoun. It means that it is my property; I bought, inherited, or created it. But when I say this is my son, my wife, or my eye I use the word *my* in quite a different sense. Day suggests that the proper meaning of that term in this context is *to appertain*. So if my wife appertains to me it means that she belongs to me, that she is a part of a certain social institution which is called marriage but by no means is she my property, and I am not her owner, though of course marriage as a social institution involves a certain set of mutual rights and obligations. In a similar sense, my eye belongs to me as a part of my body but it does not mean at all that I own it in exactly the same way as I own my glasses.

Second, Locke was wrong, because he ignored the distinction between the body-for-itself and the body-for-others. I can, indeed, as Sartre says, use the body of the other person as an instrument. But I cannot use myself. I am not an instrument and if I cannot use myself I cannot claim any right to use myself. I am not my own property. Drew Leder makes an excellent point when he remarks that "It is not that the body is like a tool, but that the tool is like a second sort of body, incorporated into and extending our corporeal powers" ([14], p. 179; see also [17], p. 217). Though I am able to move different parts of my body at will or express or imitate various moods and feelings on the scene, as it is done by professional actors, this is not evidence that I am using my body in the same way as I am using a tool. My body is simply the expression of my will and I am aware of my body because I can move it at will. The fact that I can move it and control it to a certain substantial extent, within the limits of its physiological possibilities, produces the false illusion that it is a thing, a living tool, and hence the object of ownership. If I say 'This is my body', it does not mean that this is my property. It means only that my body appertains to me, that it is integral part or the foundation of my personhood, that if I can act and choose amidst the world it is possible only through mediation of my body.

Detached from the organism the eye loses its specific function and becomes a dead thing, one of the objects of the external world, perhaps a property. Even if it were possible to sustain some rudimentary biological processes within it (and that would be necessary if the transplantation of the whole eye were available), the eye is 'dead' in the subjective sense; it does not see and it does not behold; it cannot be harmed for *there is no human subject in it*. Even if it were alive in the biological sense it would be dead spiritually. And this is the most fundamental difference between the white stick and the eye. Before I can sell or donate my living eye I have to be prepared to annihilate its 'seeing'; I have, so to speak, to kill a part of myself. No conscious life is ever present in the stick. Thus, I can make my lived body a property if and only if I die or kill myself. I can, however, make some parts of my body a property if I allow them to be removed from my body. I destroy, then, the full integrity of my body. But I cannot, logically cannot, donate or sell myself simply because I cannot own myself.

Happily, the question of the total donation of the lived body has not been discussed even by the most eager supporters of transplantation medicine. It is commonly accepted that we should prevent all donations which can cause the death of a donor. We cannot accept donation of the living heart, nor we can consent to donation of both kidneys from a healthy donor. But if I may donate one of my kidneys, why do we have such strong reservations against donating an eye? If the transplantation of the living eye is still technically impossible why I am not allowed to sell or to donate one or even two of my corneas? Before I try to answer these questions I need to say more about my relation to my body.

IV. THE HUMAN BODY

When all conscious processes in my body wither away I am dead as a human being. The dead human body is a thing, a physical body and perhaps someone's property. The experience of my body is radically different from my experience of the external world. I experience my body from *inside* – I see, hear, touch, and move it; all these acts are initiated and felt from 'inside'. The body is the boundary between the external and internal world of the human being. As the world of things is outside of me I see other human bodies (living or dead) as objects external to my body and I am perfectly aware that this is the way I am perceived by others. My body is then something unique to myself and something different for others. I see and experience myself mainly from inside; the mirror or the hand are only a secondary way of access to my own body. My inner world is accessible for others only by my mediation.

The relation between the person (soul, or mind) and the body is very tight. Some philosophers say that these are two aspects of the same entity. But I think that I am something more than the body only, or a simple sum of my mind and body. My existence in the world is total – I exist in the world as *a certain structured integral whole* and I experience that integrity in a specific way. Perhaps Roman Ingarden ([12], vol. 2, p. 197) was the first to use the concept of *primordial solidarity* to describe that relation. This is a perfect way of describing the complex phenomenon of my relation to my body. I may experience my body in many different ways. I can – as Ingarden writes – emerge in it, I can ignore it, I can withdraw from it, but I am never able to detach from it as an absolutely pure mind (soul, person); on the other hand, I cannot also identify myself in a perfect way with my body as only the body. Of course, I may imagine myself as a dead body but I cannot experience my death. The fact that my body is my body is therefore something more than identification in the world of a certain unique being; it is at the same time the expression of a certain attitude towards that being, perhaps best described by the word 'solidarity'. I do feel a sense of natural solidarity with my body because without it I would never be who or what I am – the very person I am.

(a) There are situations when my body has a distinct advantage over me. It happens when I am ill, when I lose control over my body and feel it more as a hostile or alien object than my own, friendly body [5,20]. My body hurts me then; it resists, disturbs, sometimes is a source of unbearable pain or suffering. When I feel that my body fails me I stop trusting in it, and I may sometimes treat it with some suspicion or even with aversion.

(b) There are situations where I have a distinct advantage over my body. It happens whenever I use my body apparently as a tool to achieve my ends. Yet my living body, as well as its parts, is not the tool. The tool is always a material object which is *outside* of my body; it is something which is within the scope of my power but cannot be experienced from the *inside*. Perhaps I should stress that this state of advantage over the body is a natural one which is often taken for granted. It does not matter whether I write, drive a car, play the piano, or chat with friends, I always act with and through the body, and if I have learned how to perform a certain activity I never 'think' before what should I do with my body; I am, e.g., not steering deliberately with each of my fingers while typing this essay. I think and my hands spontaneously translate my thoughts to the typed sentences. I act and use at will my fingers from inside my body. Merleau-Ponty would call it "knowledge in the hands" ([18], p. 144).

(c) These are not the only possible relations between me and my body. [I omit here the problem of the body as a conscious or unconscious means of expression of my internal life or the body as an object of acts external to it [10,21].] There is however one case which deserves to be mentioned as it aptly exemplifies that primordial state of the tension between the Ego and the body. This is the situation of the radical lack of solidarity with one's own body, when a person lives it as absolutely alien and wants its radical change. It is particularly evident in some cases of congenital diseases, or when a person does not accept his or her gender and requests sex change surgery (see [26]).

V. PRIMORDIAL SOLIDARITY

My body is a necessary condition of my being and my experience of the world. Primordial solidarity means that if my being in the world is to have any sense I have to accept my body in all its integrity and rely on it. It involves also a sort of primordial trust in my body. However, there is much more to say about the moral significance of that integrity.

(1) If I feel a sense of solidarity with my body it is not indifferent to me what happens with, within, and to my body. If I suddenly discover a black spot on my skin I fear the worst. It may indeed be a serious danger to my life. Though I may be quite ignorant about the functioning of my pancreas, liver, or bone marrow, I know that they are absolutely indispensable for the proper functioning of my healthy body, and I am seriously worried if there is something wrong with their functioning. The experienced solidarity of my body compels me to strive for maintaining its structural integrity. I cannot accept my body devoid of the eye, the kidney, the pancreas, or any other vital organ. And though it is quite

likely that I will keep on living without one eye, or one kidney, I need a really good reason to accept this kind of modification and destruction of my integrity. It seems to be quite natural to accept the loss of a kidney as result of pathological process. But the situation drastically changes if I am to consider a donation of my healthy kidney or bone marrow.

(2) I am concerned about what happens within and to my body and I have a caring attitude towards it. It is evident not only in my endeavor to sustain its structural integrity but also in my systematic 'servicing' of it. Ideally, solidarity with the body implies an attitude of responsibility for one's way of life. And the way of life is in this context the way I treat my body. If I am told by my GP that there is a deficiency of haemoglobin in my blood or an excess of cholesterol I feel obliged to undertake a positive action to restore the normal functions of my organism.

(3) I am also profoundly concerned about how other people will treat my body. Respect for my autonomy implies also respect for my body. Because I am not a thing all exclusively instrumental treating of my body may be a violation of my inherent human dignity. Nobody may touch me without my consent, and nobody has a right to intrude into my private life; although, on the other hand, it is quite likely that in some situations the good of the community may justify some restrictions to my self-determination, and violation of my privacy.

(4) Solidarity with the body implies a certain specific attitude towards other people. "The traditional philosophical way of spelling out what we mean by 'human solidarity'" – writes Richard Rorty – "is to say that there is something within each of us – our essential humanity – which resonates to the presence of the same thing in other human beings" ([19], p. 189).

I think that what Rorty calls 'human solidarity' or 'essential humanity' is based on our primordial solidarity with our bodies. If there is anything which is absolutely pertinent universally to all people it is certainly the human body. Now, if I am concerned, as I am, about what happens to my body (other things being equal) and how it is treated by other people, I am also concerned how other people's bodies are treated. Pain, suffering, disease, handicap, or torture arouse in a natural way a sense of solidarity, an attitude of caring , readiness to help or to sacrifice, or, generally, an attitude of responsibility for other human individuals. The feeling of solidarity is the stronger, the closer are our ties with the suffering person. We seem to treat it as a natural thing when somebody donates his kidney to his child or a very close relative, but we question whether a foreigner has a right to use the resources of the national bank of human tissues and organs. He is a foreigner, he is not one of us. Paradoxically, as Rorty points out, this is not the proper meaning of solidarity; all the cultural dissimilarities between people should yield to human suffering and humiliation. It is the fact that someone is suffering which is sufficient to provoke in a morally sensitive person a reaction of solidarity and the desire to help.

(5) As I said before all solidarity involves a readiness to help and to sacrifice. It may be that solidarity with other values (the life and suffering of this child) can overrule the solidarity with my body, and I may decide that it is my moral obligation or that the world will be better if I give my kidney to the dying child. But the logic of sacrifice is quite complex here. I am allowed to donate one of my paired vital organs in order to save somebody else's life; I am allowed to sacrifice my life in defence of those values with which I identify myself (e.g., in the fight for independence of my country or in striving for preserving my self-respect), but I may not sacrifice my living body for spare parts for all those in need. Why can I donate a healthy kidney when I am forbidden to donate a healthy eye, not to mention sell it? Is the single eye a more vital organ for my survival or bodily integrity than the single kidney? It is quite possible that there are different levels or degrees of primordial solidarity with the body which might be arranged along a spectrum. At one end of the spectrum one finds perhaps solidarity with the brain, then solidarity with all organs of sensual experience of the world, with eyes as the most important of all senses, and then perhaps organs of sexual identification, and all the organs essential for surviving and the proper function of the body. At the other and opposite end of the spectrum, hairs, nails. But it is also quite possible that what is at issue is not at all the functional, purely biological relevance of an organ but its social or symbolic meaning. From this point of view it might be much easier to accept a donation of some amount of brain tissue or bone marrow than the scalp, or a breast.

(6) Because I feel solidarity with my body I am also concerned about what will happen with my dead body. It does not mean however that I feel solidarity with my corpse. You can feel solidarity only with sentient beings who are able to feel the same sort of emotions as you. One cannot have any sense of solidarity with things. My dead body is a thing but at the same time it is something more than a common thing. It is a symbol of my former presence in the world, and though at present I may have some specific requests with respect to the treatment of my body after my death, which of those requests will be respected is determined straightfor-

wardly not by my wishes, desires, or interests but by the norms of my culture. If I claim to have a right to have my dead body treated with respect the meaning and the manner of showing that respect is determined by the tradition of the society I happened to live in. It is quite likely that if I had lived in a primitive tribe some hundred years ago my liver would be eaten by my relatives or the chief of the tribe; as likely as that some of my tissues and organs would go for transplants if I had happened to live and suddenly die in a car crash in this country having a donor card in my pocket. But it does not mean, however, that my genuine wish to have my dead body converted into pet food and used for feeding homeless dogs and cats will be seriously treated by my relatives. The moral tradition and the law of the country decide what sort of actions are proper or improper with respect to dead persons. But as a community we do feel a sense of solidarity in the way we treat the bodies of the dead. Otherwise it might be quite difficult to explain the universal occurrence of funeral rites.

VI. GIFT VERSUS COMMODITY

It is not difficult to explain the origin of the controversy over ownership of the human body. Technological progress in medical sciences and the realization that all products, cells, tissues, and parts of the human body may have a substantial market value have caused a situation where both the patients as potential beneficiaries and the medical establishment as providers would like to find some ideological justification for available ways of diagnosis and treatment. However, the main complication of many discussions concerning the moral and the legal status of the human body is the fact that one tries to find *the* universal answer to three essentially different questions at the same time: (a) who owns my living body?; (b) who owns the organs, tissues, cells and products generated by my body?; (c) who owns my dead body?

I, for one, deeply believe that question (a) is the wrong question to ask. Neither I, nor my parents, nor the society I belong to, nor God (if existing) own my living body. As a person who is a certain structured integral whole, who cannot even imagine himself without a body, I see no reason to view my living body as a form of property. My relation to it is the relation of solidarity with it, but certainly it is not the relation between the proprietor and property.

But even if we accept that the living body is not a form of property it does not solve at all the problem of the ownership of all organs, tissues, and products generated by the living body ([1,2,3,4,6,8]). If they are entirely detached from my body, and are now physical objects outside of my body, it seems that I may decide what I should do with them. I may donate them but I may also sell them. The language of gift is the language of solidarity. The language of selling, the language of commodity, is the language of market economy. At present we seem to be at a crossroad, because if we are to establish the proper law we would like to reconcile the language of solidarity with the language of utility, or market economy. I am afraid it is an impossible task. If we choose the language of solidarity it means that I and only I have the right to decide how my organs, tissues, or cell will be used. I may or may not donate it. But nobody has any right to claim my kidney or sue for a legal injunction to give my bone marrow to my cousin [3]. If we choose the language of property then some people may make me offers I cannot refuse. We could then of course save more human lives and greatly reduce the amount of suffering but the moral price for it seems to be too high to pay. My body would then be treated as a sort of spare parts store, or a biological substance, too precious to be wasted.

The ethics of solidarity may also suggest an interesting solution to the status of dead bodies. One may argue that human solidarity with suffering people is a strong reason to feel morally obliged to make gifts of our dead bodies to them. It is also very likely that the ethics of solidarity provide the best moral justification of a presumed consent policy. Granting to the individual the right to opt-out we may nevertheless try to educate people to see and feel solidarity with the suffering of other people. The best example of such an educational policy is perhaps the donor cards program. If we had decided that body parts have above all a utility value and, as a commodity, are the objects of trade, we shall ignore the suffering of the potential recipients, giving priority not to moral, but to economic values. (Should my family, e.g., sell my dead body? How much can they make? Will it be enough to pay the college fees of my son?)

One thing is certain, the language of solidarity, although it is not perhaps able to solve all practical conflicts resulting from treating human organs, seems to be morally more acceptable in the context of medicine than the language of property. The ethics of solidarity is very close to the ethics of reciprocity. It is not wrong to appeal for a gift of an organ for a suffering baby, but it is morally repugnant to many to make a business out of human suffering.

VII. A GIFT OF SIGHT

Suppose for a while that we have accepted the language of solidarity and put a total ban on all commercial transactions in solid organs procured from living donors. Could I then, acting from a pure and disinterested altruism, make a gift of my healthy eyes to those in need? It seems to be an odd question and some people certainly would advise me to contact a psychiatrist. But, as a matter of fact it is not an inconceivable case. If we are full of admiration for a mother who has decided to give her healthy kidney to her sick son, what should we think about a father who is ready to sacrifice his own cornea to prevent imminent blindness of his only and beloved daughter? Why, if we accept the gift of a kidney, do we hesitate in accepting the gift of an eye?

The general principle, which may explain our hesitation, is simple and evident: we should not accept donation of those tissues and solid organs, procurement of which may cause the death or grievous bodily harm (mutilation) of the donor. The problem is that the dynamic development of transplantation medicine can bring about a further revision of the concept of mutilation or grievous bodily harm. For many surgeons of an earlier generation the removal of a kidney from a healthy donor was the act of inflicting grievous bodily harm, absolutely incompatible with the Hippocratic Oath. One can read nowadays about surgeons who see nothing wrong in mastectomy of healthy breasts, if such is the wish of the patient [26]. It may be that in order to facilitate procurement of transplantation material our concept of mutilation will undergo further erosion. Generally, we do not see anything wrong in procurement of those tissues and organs the removal of which does not change the bodily integrity of the patient, where the bodily integrity of the patient is understood by what is outside but not inside of the body. In social perception a kidney, blood, or bone marrow donor does not differ at all from non-donating individuals. But the evident destruction of the visible body usually provokes quite a strong emotion. Only recently has the mass-media informed us about transplantation of a post mortem recovered face. It is supposed to be an effective treatment for extended and severe burns of the face. It is difficult not to be frightened by this form of 'therapy'. We are used to thinking that the face, and particularly the eyes, are the integral elements of our identity and that the very possibility of transplantation of these part of the human body furnishes a real threat to our sense of personal identity.

Transplantation medicine has changed our feeling of what is permissible and not permissible to do with and to our bodies. The most uncompromising enemies of transplantation programs claim that we have stepped on the slippery slope leading to a bizarre, modern form of cannibalism. It is not clear to me at what point we should draw the line, but I have no doubts that even human solidarity has its limits.

Centre for Philosophy and Health Care
University of Wales Swansea
United Kingdom

ACKNOWLEDGEMENTS

I would like to express my deep gratitude to my colleagues from the Dutch project on the ownership of the body, particularly professors Henk ten Have and Stuart F. Spicker for their friendly comments and helpful inspiration. I wish also to thank my colleagues from the Centre for Philosophy and Health Care in Swansea, particularly H.M. Evans, D.A. Greaves, and H. Upton, for their constructive criticism and linguistic assistance.

BIBLIOGRAPHY

1. Andrews, L.B.: 1986, 'My Body, My Property', *Hastings Center Report* **16**(5), 28-38.
2. Andrews, L.B.: 1989, 'Control and compensation: Laws governing extracorporeal generative materials', *Journal of Medicine and Philosophy* **14**, 541-569.
3. Calabresi, G.: 1991, 'Do we Own our Bodies?', *Health Matrix. Journal of Law-Medicine* **1**(5), 5-18.
4. Campbell, C.S.: 1992, 'Body, Self, and the Property Paradigm', *Hastings Center Report* **22**(5), 34-42.
5. Cassell, E.: 1992, 'The Body of the Future', in D. Leder (ed.), *The Body in Medical Thought and Practice*. Kluwer Academic Publishers, Dordrecht/Boston/London, pp. 233-249.
6. Cohen, L.R.: 1991, 'The Ethical Virtues of a Futures Market in Cadaveric Organs', in W. Land, J.B. Dossetor (eds.), *Organ Replacement Therapy: Ethics, Justice and Commerce*. Springer-Verlag, Berlin/Heidelberg, pp. 302-310.
7. Day, J.P.: 1966, 'Locke on Property', *Philosophical Quarterly* **16**, 207-220.

8. Dickens, B.M.: 1991, 'Who Legally Owns and Controls Human Organs After Procurement?' in W. Land, J.B. Dossetor (eds.), *Organ Replacement Therapy: Ethics, Justice and Commerce*. Springer-Verlag, Berlin/Heidelberg, pp. 385-392.
9. Feinberg, J.: 1980, 'The Interest in Liberty on the Scales', in *Rights, Justice, and the Bounds of Liberty*. Princeton University Press, Princeton, NJ, pp. 30-44.
10. Harré, R.: 1991, *Physical Being*. Blackwell, Oxford, U.K.
11. Honore, A.M.: 1991, 'Ownership', in A.G. Guest (ed.), *Oxford Essays in Jurisprudence*. Oxford University Press, Oxford, U.K., pp. 107-147.
12. Ingarden, R.: 1960, *Spór o istnienie swiata* [*The Controversy over the Existence of the World*], (2nd ed.) PWN, Warszawa. The German version: *Der Streit um der Existenz der Welt*, (3 vols.) Niemeyer, Tübingen, 1965.
13. Jonas, H.: 1970,'The Nobility of Sight: A Study in the Phenomenology of Senses', in S.F. Spicker (ed.): 1970, *The Philosophy of the Body. Rejections of Cartesian Dualism*. Quadrangle Books, Chicago, pp. 312-333.
14. Leder, D.: 1990, *The Absent Body*. University of Chicago Press. Chicago/London.
15. Leder, D. (ed.): 1992, *The Body in Medical Thought and Practice*. Kluwer Academic Publishers, Dordrecht/Boston/London.
16. Locke, J.: 1690, *The Second Treatise of Government*, chapt. 5, sec. 27.
17. Marcel, G.: 1970, 'Metaphysical Journal and A Metaphysical Diary', in S.F. Spicker (ed.): 1970, *The Philosophy of the Body. Rejections of Cartesian Dualism*. Quadrangle Books, Chicago, pp. 187-217.
18. Merleau-Ponty, M.: 1962, *Phenomenology of Perception*, trans. Colin Smith. Routledge and Kegan Paul, London and Henley, U.K.
19. Rorty, R.: 1989, 'Solidarity', in *Contingency, Irony, and Solidarity*. Cambridge University Press, Cambridge, U.K., pp. 189-198.
20. Sacks, O.: 1984, *A Leg to Stand On*. Picador, London, U.K.
21. Scarry, E.: 1985, *The Body in Pain*. Oxford University Press, New York.
22. Sartre, J.P.: 1970, 'The Body', in S.F. Spicker (ed.): 1970, *The Philosophy of the Body. Rejections of Cartesian Dualism*. Quadrangle Books, Chicago, Il., pp. 218-240.
23. Snare, F.: 1972, 'The Concept of Property', *American Philosophical Quarterly* **9**(2), 200-206.
24. Spicker, S.F. (ed.): 1970, *The Philosophy of the Body. Rejections of Cartesian Dualism.* Quadrangle Books, Chicago, Il.
25. Straus, E.W.: 1970, 'Born to see, bound to behold: Reflections on the function of upright posture in the esthetic attitude', in S.F. Spicker (ed.): 1970, *The Philosophy of the Body. Rejections of Cartesian Dualism*. Quadrangle Books, Chicago, Il., pp. 334-361
26. Stuart, A.: 1992, 'I am living proof of the third sex', *The Independent* (12 May).

PART III

MEDICAL INTERVENTIONS AND STATUTORY FOUNDATIONS

JOS V.M. WELIE
HENK A.M.J. TEN HAVE

OWNERSHIP OF THE HUMAN BODY: THE DUTCH CONTEXT

I. INTRODUCTION

Ownership is a well-known feature of the world; without it our world would be considerably different. Consequently, ownership is an important topic in moral, political, and legal philosophy. Furthermore, every country has its particular ownership institutions. All too human quarrels about what's mine and what's thine have led to the establishment of an extensive body of laws. Intricate legal distinctions have been developed between, for example, the actual power to dispose of goods (possession), the title or legal right to dispose of goods (ownership), and the goods being owned (property). This chapter intends to investigate the issue of ownership of the human body and body parts from a Dutch legal perspective. However, nowhere in the Civil Code of The Netherlands is anything said about ownership of the human body and body parts. Houses, roads, trees, animals, corn, even treasures are mentioned in property law, but nothing about human bodies or body parts.

Our intuitions seem to fluctuate between two extremes, and so does the law. On the one hand, we tend to consider the body to be a 'thing' like any other. Since the person incorporating his body can be thought to 'hold' or 'possess' it, he may be assumed to be the owner of his body as well (art. 107 Book 3 Civil Code j° art. 1 j° art 4 Book 5 Civil Code). An owner can decide what to do with his things, his property (art. 1, Book 5 Civil Code). Indeed, the human being can decide to undergo or refuse medical treatment of his body, decide to donate blood or sperm, or have his hair cut and even sell it. Acts, such as *donation* and *sale*, are regulated in detail by civil law; medical treatment is also addressed by the Civil Code. Not only may a person decide (by means of a testament) what is to be done with his property after his death; he can also decide what is to be done with his corpse.

On the other hand, society seems unwilling to allow a person to do with his (dead) body whatever (s)he wishes (one is reminded of Leenen's

H.A.M.J. ten Have and J.V.M. Welie (eds.), Ownership of the Human Body, 99–114.

thought-provoking example of donating one's dead body to be fed to animals in a zoo [7]). Furthermore, if parents of minors are the legal owners of all of their children's possessions (as they are under Dutch law), and if the human body is a piece of property then parents would own their children's bodies. If society hesitates to allow an adult person to donate his kidneys while living, surely parents may not donate the organs of their living children. Here we encounter an obvious legal limit to the 'body=property' theory.

Yet body parts have always been legally owned by someone. Body parts have been sold, donated, or borrowed, that is, they have been manipulated in accordance with the guidelines provided in the Civil Code. Human bones and skeletons continue to be purchased by schools or donated by archaeologists to museums of anthropology. However, recent technological developments have brought the issue much closer to home. Nowadays people are donating blood or kidneys, selling sperm, and research laboratories are purchasing cell lines from living patients. The organs and tissues taken out during simple surgical interventions (which patients believe are being destroyed) may end up yielding an enormous profit when sold to and processed by a pharmaceutical company. With the recent advances in transplantation surgery, virtually all organs and tissues of a deceased human being are re-useable, while advances in human genetics will soon enable the procurement of interesting information about an individual from a single (donated) white blood cell.

All such actions are considered relevant under Dutch civil law. Although lawyers have been debating the legal status of the human body and body parts for many decades and complaining about the vagueness of the civil law in this regard (for an old example, see [14]), the new Dutch Civil Code (1992) only contains one article pertaining explicitly to human body parts (art. 467, to be discussed hereunder). This fact allows for two contrasting conclusions: either the human body and its parts are in no way different from cars, apples, dogs, and any other property item; or the human body and its parts are so different from these items that they cannot be owned at all; that is, they fall outside the scope of property law.

The history of Dutch jurisprudence is equally silent regarding this issue ([4] p. 155). Although in 1946, the Dutch Supreme Court decided that removing the artificial teeth of a corpse (without consent) is 'theft' (as defined by article 310 of the Criminal Code), this does not imply that the body or its parts are property under the civil law. Article 310 of the Criminal Code defines theft as taking away a good that 'belongs' to another person, not that it is 'owned' by another person. Some authors have argued that the Supreme Court's use of the term 'apprehensive avoidance' instead of the term 'property' or 'ownership' implies that the Court considers the human corpse not to be a piece of property ([18], p. 299). But others have pointed out that the Court did not decide this matter and that, consequently, no inference concerning the Supreme Court's opinion can be drawn [4,22].

Thus, the only way that remains for approaching the issue of ownership of the human body is to begin with the concept of ownership itself. As is the case with virtually all fundamental concepts, ownership is not defined in Dutch civil law. Therefore, we can but explore the various ways in which the concept of ownership is interpreted and employed in the Dutch Civil Code and jurisprudence. First, we shall examine the issue of ownership of the entire living body. Obviously, if a living human body can be owned, it is only one's own body. Owning somebody else's body is more problematic. If we can establish that every person (legally) owns his own body, the issue of ownership of one's body parts is resolved. For according to the Dutch Civil Code, ownership of an item implies ownership of the parts that constitute the whole, as well as the 'fruits' generated by the whole. We will therefore address the issue of ownership of the entire living body first and then consider the issue of ownership of body parts: organs, tissues and the corpse.

II. THE CONCEPT OF OWNERSHIP

As mentioned earlier, the Dutch Civil Code does not define the concept of ownership. Article 1 of the 5th book of the new Dutch Civil Code merely states that ownership is the most comprehensive right which a person can have to a thing. Three aspects of this proposition stand out: (1) ownership is a *right of a person to a thing*, (2) ownership pertains to *things* only, and (3) ownership is the *most comprehensive right*, but this right is not unlimited. Hence, the first question we have to address is whether the human body can be the object of a legal relationship. Every legal relationship has two poles, the subject and object of the relationship: someone has a right to something. Although the Civil Code does not mention the human body as a possible object in a legal relationship, the Constitution and other Dutch laws seem to suggest otherwise.

III. THE STATUS OF THE HUMAN BODY IN THE DUTCH CONSTITUTION

When the Dutch Constitution was completely revised in 1983 the human body was granted special protection. Article 11 of the Constitution states:

> Everyone has a right to untouchability of his body, except for restrictions provided by or valid because of the law.

This article suggests that the human body can be an independent object of a legal relationship. Before this conclusion can be drawn the article must be examined more closely. Three aspects of article 11 are in need of further discussion: (1) the body to be touchable, (2) the untouchability is not restricted to harm coming from third persons, and (3) the untouchability is not absolute, for it can be limited by formal law.

(1) The term 'untouchable' is interesting as well as confusing. Akkermans questions why this term was chosen instead of a more commonly used term such as 'integrity'. Whereas *un*touchability is a negative term in that it defines what is *not* permitted, the positive term 'integrity' indicates what deserves protection without prescribing how that protection is to be provided. When, for example, in discussions about abortion it is argued that a woman has a right to the integrity of her body, the point is not intended to keep the woman from 'touching' her pregnant body, but exactly the opposite such that she may undo the violation of her bodily integrity. On the other hand, the fact that the legislature apparently preferred the term 'untouchable' suggests that it intended to stress the point that the human body may not be harmed by anyone. This interpretation is consistent with the implications of the second aspect of the article [1].

(2) The article suggests that the protection of the body is not restricted by the freedom of the bearer of the right. The article does not say that the body may not be touched by others. The body is untouchable for anybody, including, therefore, the person whose body it is. The right of self-determination, to freely decide about one's own course of life, does not include the freedom to use one's body in any desired manner. For example, this constitutional right cannot be equated with the maxim of freedom-of-abortion advocates that women are 'boss over their own belly'. Apparently the special ethical status of the human body is not deduced from the general right to freedom of the human person, for it actually limits that freedom. The special ethical status is integral to the

human body itself and the body deserves special protection, irrespective of the wishes of the person incorporating that body. Any contract that may obligate the underwriter to undergo bodily harm, even if such harm serves as a punishment, is void. Similarly, a contract about the prospective sale of separated body parts (e.g., hair, teeth, sperm) cannot be enforced if the seller decides not to cut his or her hair, or to separate any particular body part from the body as such. The contracted buyer may demand compensation for prior expenses incurred, but he cannot demand the delivery of body parts previously negotiated [22].

These considerations suggest that the inclusion of article 11 in the 1983 Constitution was motivated by the traditional canon that the human body is a sacred good lent to man by God; or if not lent by God personally, then at least of a 'higher order' than the order of common inorganic and inanimate objects, and hence an object of intrinsic dignity. But the Dutch parliamentary debates preceding the new Constitution show differently. They reveal that the legislature may not at all have intended to limit the freedom of choice of the owner of the body in view of the body's intrinsic value and dignity [20]. Initially, the legislature did not even consider including in the Constitution an article directed to the protection of the human body. It was thought that article 10, which safeguards personal privacy, would guarantee protection of the entire human person including his body. Obviously, privacy is first and foremost a matter of individual choice: protecting someone's privacy in a manner that overrides the very wishes of that person, is incoherent. Parliament finally decided to insert article 11 when rumors and reports surfaced alleging that research was being conducted on former detainees, that all kinds of mandatory medical treatments were being advocated (such as fluoridation of water, involuntary castration, sterilization and electroshock therapy), that in neighboring countries terrorists were being tortured in prisons, and that the ethical issues concerning transplantation of organs were ever more pressing. However, parliament was not suggesting that the body needs protection over and beyond the right to protection of personal privacy. The human body was considered to be protected equally by articles 10 and 11 [1].

Since 1975, victims of battery can request financial relief from a special state foundation. It has been argued that this demonstrates that the State considers the human body such an important good that in addition to assigning a liberal right (to protection against harm), the State acknowledges a positive right (to financial assistance) [20] (cf. the liberal right to freedom of education and the positive right to state subsidies for education). However, one can also speculate that it is the serious suffering that resulted from the assault that merits such special financial assistance, rather than the physical harm to the body.

The aforementioned examples of offenses against the body underscore the fact that parliament was mostly concerned about the danger to the human body posed by outsiders (e.g., cutting hair, treatments by a physician, drawing blood, vaccinations, mandatory feeding, pumping a stomach, medical experiments, and physical punishments). When the question was explicitly raised whether art. 11 implies that the human body was to be protected against its owner as well, the government responded that one of the elements of the right to 'untouchability of the body' is the very right of its owner to control it. While this answer invokes a potential paradox, the issue was not further explored by the legislature. It remains unclear how much control the legislature was willing to grant the owner of the body. Abortion, suicide, and euthanasia were debated but final conclusions were not reached.

Given the greater lenience of Dutch society toward suicide and euthanasia, actions which result in death *without* anyone else but the deceased benefitting, it is remarkable that there remains a considerable and apparently undiminished opposition to *altruistic* self-sacrifice. If a person decides to donate both of his kidneys or his liver to someone else, he probably will not be able to do so. Altruistic acts with lethal consequences are frequently deemed contrary to morality and law; few surgeons seem willing to participate in such actions ([22], p. 311).

In any case, article 11 of the Constitution was not intended to bestow on the human body a special independent legal status worthy of legal protection, but rather to indicate that a person's physical nature was as much a matter of privacy as his mail, thoughts, or choice of life-style. That is not to say that the human body is not protected against harm. The criminal law contains various articles that specifically prohibit bodily assault (articles 300-306). But again it is not clear whether these articles imply that the human body is an *independent* item of moral value worthy of special legal protection. As Leenen has pointed out, when harm is done to the human body, the offender is not charged with damage to property, but assault of a human being ([7], p.4). For similar reasons, Van Bemmelen and Van Hattum have argued that assault is a criminal offence because such an act endangers the life of the person ([2], p. 209). That would also explain why assault is not justified by a person's request to be harmed: human life is of such a high value that killing on the victim's request, assisting in his suicide, and even any bodily harm that serves no other purpose but to harm and thus to endanger life, is illegal in The Netherlands ([9], p. 971).

(3) The conclusion to the preceding analysis of the second aspect of article 11 seems to be that the protection of the physical body noted in the Constitution and the Criminal Code is *not primarily the result of some independent value of the human body*, but first and foremost the logical consequence of the value assigned to personal self-determination and to human life (see also [3], p. 188). To verify this conclusion, we shall examine the third aspect of article 11: the body may be legally "touched", but only if such infliction is allowed for by a formal law. A typical example of such a legal violation of the untouchability of the human body is the examination of the body of a suspect to acquire trespasser for criminal evidence. Indeed, during the parliamentary debates on the new constitutional article 11, body searches and drawing blood were explicitly mentioned as actions that could injure the integrity of the human body ([20], p. 4).

Currently, Dutch criminal law allows arrested people to be 'examined on body and cloths' (art. 56). The phrase '*on* body and cloths' (*aan lichaam en kleding*) indicates that only an examination of the surface of the body is permitted. Cutting or pulling hair (not collecting hair), drawing blood and obtaining sperm, and taking X-rays are not warranted under article 56 of the Criminal Process Code ([19], p. 2; [15]). Examination of the natural openings of the body (such as the mouth, nose, anus, or vagina) is permitted [5], but sampling of some cells from the inner cheek (endothelium) for laboratory research would not be allowed [16].

These two limitations clarify that article 56 (which, by the way, is older than art. 11 of the Constitution) does not constitute a genuine exception to article 11 of the Constitution. No real harm may be done to the body: no incisions may be made, no needle may be injected into the body. The physical examination of the suspect is noted in the same article as the examination of the clothes, because the body is not supposed to be examined any more thoroughly than are the clothes. In essence, so-called 'frisking' is permitted, and so is being sniffed at by a dog [6] so long as one is searching for traces of a crime, not searching for personal information about the suspect. For the very same reason, the second limit is set: examination of the surface of the body, including natural openings of the body, is intended to find objects that are not part of the human body, whereas the purpose of sampling cells for laboratory examination is to obtain information about the identity of the person himself (e.g., DNA-fingerprinting) ([19], p. 2). In addition, there is no legal obligation to cooperate in such an examination of the surface of the body, but the arrested person is not allowed to obstruct such an examination. This means that a person does not have to open his mouth when he is ordered to do so, but he must tolerate that his mouth will be forcefully opened! ([2], p. 119-120).

There is one exception to the rulé never to actually harm the body. Article 26 of the Traffic Law of 1974 allows drawing blood to test for drunken drivers. But the blood test for drunken driving cannot be forced upon the suspected driver. He may refuse to have his blood drawn. However, is he refuses he will simply be assumed to be intoxicated; in that sense, some coercion is exerted on the drunken driver to participate in the test. Obviously, the blood tested for alcohol may not be used to test for any other substance ([17], p. 281)

It should be mentioned that currently there is a broadening of the limits on obtaining criminal evidence, particularly in the area of DNA-fingerprinting. When, in 1989, the District Court of Maastricht refused to grant the prosecutor the right to undertake the DNA-examination of a suspect, the Minister of Justice announced that there was a need to formulate a law to allow for a 'violation' of the untouchability of the human body for such DNA-examination ([21], p. 173-174; [19], p. 8). In answer to this call by the Minister of Justice, the State Commission for the Evaluation of the Criminal Process Code created a subcommission to prepare such a bill. Reijntjes suggested that the Commission should not make a special law only for DNA-examinations. After all, there is no difference in principle between comparing DNA-fingerprints or undertaking the more traditional comparison of blood proteins. The DNA-fingerprint comparison may be more reliable, but as long as the blood is not used for other purposes than to determine who committed the crime at hand, there is no principle difference. Blood taken for alcohol testing may not be used for other purposes either ([17], p. 281).

To recapitulate: article 11 of the Dutch Constitution does not guarantee the integrity of the body on the basis of an assumed, inherent value of the human body, but rather addresses the decision-making power of the person over his own body. Similarly, the Criminal Code punishes physical assault because such assault inflicts suffering on the person and endangers his life. Finally, the Criminal Process Code provides no reasons to infer that in Dutch law the human body is respected independently of the person whose body it is. Consequently, there are no indications that the body can be an independent object of any legal relationships.

Obviously, lack of arguments in favor of, in and by itself is no argument against. But as Leenen has pointed out, arguments can be proffered against the idea of a person owning his body. For such a position implies a dualistic anthropology: if something cannot own itself, the owner of the human body cannot be the same human body. The human being is both (1) a non-bodily person who is the bearer of various rights, such as the right to property, and owner of the body, and (2) the body which is being owned. Leenen strongly objects to the theory of a non-bodily person as 'owner' of the body. He contends that from a legal perspective it is not possible to construe a bearer of human rights that does not have a body. A person acts in society by virtue of his body and, consequently, the bodily person is the bearer of rights [7].

We conclude that from a Dutch legal perspective the human body cannot be considered an object of *legal* relationships such as ownership. The situation changes, however, when elements or parts of the body are separated from the body (e.g., donated blood or amputated leg). Those parts no longer can be considered integral to the bodily person. Consequently, the very reason for disqualifying the entire human body as a thing, a piece of property, ceases to exist. Separated organs may be objects of legal relations, including ownership. We therefore shall initiate the examination of the three crucial aspects of the legal delineation of ownership anew: (1) ownership is a *right of a person to an object*: body parts must be considered independent objects; (2) ownership pertains to *things* only: body parts must be considered things in the sense of the law; and (3) ownership is a most comprehensive but not necessarily unlimited right: the legal formula which describes the relationship persons have towards human body parts is ownership. As we will see, there exists considerable disagreement notably about the latter issue. To settle the disputes, we will have to closely follow the law itself in order to obtain evidence in favor of or against any of the positions mentioned above.

IV. OWNERSHIP OF BODY PARTS

There is no doubt that human body parts can be independent objects of a legal relationship. As was mentioned earlier, people *donate* blood and organs, *purchase* bones, *lend* a skull, *sell* a cell-line. There is, however, no consensus on the nature of these relationships.

To be owned, a body part must be considered a 'thing' (*zaak*). Article 1 of book III of the Civil Code defines 'things' as material objects that can be manipulated. Whether the human body fits this definition is not explicated, nor can it be derived from any other article in the same section of the Code. But it seems obvious that a body part is material and can be manipulated. Nevertheless, the question remains whether *all* manipulable objects are things, whether, in other words, human body parts are *mere* things. Whereas some authors seem to consider body parts things like other things, others are not willing to accept the philosophical consequences of such a position.

Leenen claims that from the moment of separation, body parts can be considered things that are the property of the person from whom they were separated, until (s)he has donated, sold, or disposed of those parts ([7], p. 5). Although many authors agree with Leenen that body parts when separated from the entire body become things owned by the person from whose body they were separated (e.g., [4], p. 157; [13], p. 431), there is less clarity about what happens afterwards. A barber is obligated to return the client's hair to the client if the latter so desires. The surgeon removing gallstones is required to do the same. But what about blood donated to the blood bank, or the residue of urine from a pregnant woman originally sampled for general diagnostic tests? Both are potentially of value to the recipient and he may not want to return them to the donor. The former, for instance, could be used for biomedical research, the latter for the commercial production of drugs.

Leenen contends that organs donated to a recipient change owners [8]. This is not the view of the National Commission for the Fight against AIDS (Nederlandse Commissie AIDS Bestrijding). The Commission has argued that the blood of a donor, even after being donated, remains the *donor's property* if the blood bank intends to use it for purposes other than those of the donor. For example, donated blood is the property of the blood bank when the blood bank uses it for transfusions, but not when the bank intends to use it for research into the prevalence of AIDS in the Dutch population, even if the bank guarantees the confidentiality of the donors. The blood bank must ask permission from the donor since the donor remains (partial) owner of the blood ([9], p. 5). Given that the legal ramifications of ownership imply the freedom to decide about its use (while that use is restricted only by (un)written legal rules), the Commission's position is rather unusual. Ownership can be shared but not *split* as the Commission seems to advocate.

Undoubtedly, the opposition of the Commission is not motivated by a desire to guarantee that the original donors of blood and other body parts will share in the financial benefits of pharmaceutical profits obtained through commercial processing of these tissues. The Commission is more concerned about the protection of the confidentiality of the donors. Since human tissues, even single cells, contain a host of identifying information, ranging from blood type to phenotype, the constitutional right to privacy prohibits the recipients of body parts that are not separated from their own body, from using these parts in just any desired manner. The power of control based on the right to protection of one's privacy, may surpass the power of control based on ownership. In this sense, body parts are not different from the paper in the patient's hospital's file: the hospital owns the paper, but since the information written on it pertains to the privacy of the patient, the hospital may not use its property in just any desired manner.

To guarantee the confidentiality of the original donor, it seems more in accordance with the Dutch legal system to argue that the recipient of a body part becomes a full owner but his ownership is conditional: if the body part is used in any other way than provided by the former owner, the recipient automatically looses his property and the former owner becomes the owner once again. Olsthoorn-Heim has suggested a similar approach [11]. She maintains that blood and body parts that are extracted or donated remain the property of the donor, even though the hospital or blood bank possesses them de facto. Ownership by the patient or donor implies full decision-making control. Consequently, the medical institution has to obtain the informed consent from the donor; consent may be *presumed* only for procedures with which the patient or donor is likely to be familiar (such as the cremation of an amputated leg or the transfusion of donated blood). Article 467 of the new civil law section on health care contracts specifies that consent may be presumed for research with anonymized body tissues, but only if the donor's continued anonymity is guaranteed and if, obviously, the patient has not objected to this research.

V. OWNERSHIP OF THE CORPSE

The ownership issue becomes even more complicated when, instead of organ or tissue ownership, the status of the human corpse is to be assessed. Leenen's theory regarding the separation of body parts from the living unity, turning the separated body part into a thing, could be applied to the corpse as well. The corpse, too, has become a thing at the moment of final disintegration of the wholeness of the human person, that is, at the moment of death. Nonetheless, many authors distinguish between parts separated from the living body on the one hand, and the corpse on the other, the dissimilarity being that the corpse is a whole. Indeed, it has been argued that the untouchability of the living body extends to the dead body ([1], p. 250). Yet unlike desecration of graves, desecration of corpses is not a criminal act in The Netherlands (Article 149 Criminal Code) ([9], p. 295). Destruction or dissection of the corpse without governmental permission is punishable (Article 41 Law on the Disposal of the Dead). But the principle reason underlying this prohibition is of a different nature: correct disposition of the dead body is a matter of public importance which requires governmental supervision.

Oven maintains that a detached body part is, indeed, a thing owned by the person from whom it was detached, but the dead body is not owned by its former inhabitant. That is not to say that it is a 'res nullius', an object that is not owned by anybody but may become a piece of property when somebody appropriates it. Oven calls the dead body a 'res extra patrimonium', a thing outside the range of property rights ([12], p. 18-19). Oven points out that in Roman law, the church yard was not an object of trade and so, he concludes, *a fortiori* the buried corpses could not be objects of trade.

Wery, however, contends that the corpse cannot be left in such a 'legal vacuum'. The dead body must be owned by the formerly alive person. The only other alternative that the corpse becomes the property of the heirs Wery rejects as too bizarre ([22], p. 306). But Wery's solution is problematic too, for a non-existent person cannot truly be said to own something. Indeed, the testament (will) of a former person may still have binding power, but that, one might contend, is not because of the rights of the non-existent person, but rather the implied promise by society to honor somebody's wishes, even after that person was deceased. Not the theoretical construction of a non-existent bearer of rights but the obligatory nature of promises justifies the legal force of testaments (wills). The property formerly owned by the deceased person no longer is his property; it has become the property of the heirs upon his death. The testament merely regulates who among the heirs is the rightful owner of which things.

Leenen provides another reason for considering the corpse of a deceased person a material thing and the legal property of the heirs. If the bones in an anatomy museum, the mummy in an archeological collection, the skull on an anthropologist's desk, or the skeleton in a biology class are owned by the museum, the collector, the anthropologist and the school respectively, while all these body parts originally were owned by the deceased person, how is it that they changed rightful owners? It would be fully contingent, to indicate the moment in the decomposition of the corpse at which the deceased person ceased to be the owner of the corpse ([7], p. 7). Only if the deceased person, while being alive, donated those parts, could they have changed owners. But most bones end up in museums because the old graves on the church yards were removed. Egyptian pharaohs certainly never donated their bodies to the museums.

Wery tries to solve Leenen's 'contingency problem' by arguing that the dead body must be considered the property of the deceased as long as this is required by the interests of the deceased person. Wery notes that the Law on the Disposal of the Dead grants the right to decide about the disposal of the corpse to the surviving relatives of the deceased, not to his or her heirs. After all, relatives are usually the most reliable source for determining what is in the best interest of the deceased when an explicit testament is unavailable. But Wery's solution to Leenen's 'contingency problem' is rather contingent itself. For it clearly would have been in the best interest of Egyptian pharaohs – given their religious convictions – to have considered the mummy the property of the pharaoh rather than of the museum. The 'as-long-as-this-is-required-by-the-interests-of-the-deceased-person' construction may range from a few days to some 4,000 years. In fact, Wery's criteria for the 'as-long-as-this-is-required-by-the-interests-of-the-deceased-person' standard seem to be more germane to the survivors than the deceased: the recognizability of the corpse, the fact that people who knew the deceased person are still alive, or their understanding of the purpose of using the corpse ([22], pp. 306-308).

Leenen concludes that the corpse of a deceased person should be considered the property of his heirs ([7], p. 7). That is not to say that any manipulation of the corpse is allowed. We already mentioned the testament of the deceased person that may state what is to be done with the corpse. Various other laws, such as the Law on the Disposal of the Dead, further limit manipulation of human corpses. Finally, common standards of decency and proper customs prohibit arbitrary manipulations. As Van Praag, another advocate of the body=property theory, has pointed out, there are many property items that cannot be legally used in any desired manner by the owner. For example, crown jewels may not be sold by the royal owners, until the royal family is exiled from the country; church properties may not be sold for profit unless and until the Church itself is liquidated ([14], p. 96).

Thus, the mere fact that the Law on the Disposal of the Dead, in the absence of a living will, grants the next of kin the right to decide whether the corpse may be used for transplantation (articles 72 and 77), rather than the legal heirs, is not a sufficient argument against the view that the heirs own the corpse. Moreover, the expenses of the disposal are to be paid from the heritage, which is owned by the heirs (art. 288, Book III, Civil Code). Ownership and the power of control are related, yet different rights. The heirs may own the dead body, but the formerly living person, either by means of a living will, or by testament to the surviving next of kin, retains the power of control, including the power to donate his corpse to a medical school (which then becomes the owner), or to have his corpse buried (which would make the corpse a 'res nullius', possibly exhumed and appropriated in 2,000 years by archaeologists).

While the body=property approach may be more consistent with regard to the dead body, it gives rise to a rather paradoxical situation: since (1) organs donated for transplantation become the 'property' of the recipient, (2) a corpse donated for medical research and education becomes the property of the medical school, and (3) a corpse which is buried or burned becomes a res nullius, and since the Law on the Disposal of the Dead does not allow for any other manner of disposal (art. 1), the heirs never really have the opportunity to obtain their inheritance.

VI. CONCLUSION

The discussion of the legal ramifications of ownership of the body reveals that there exists no consensus in Dutch theory of law, except for one notion: in principle, the person who embodies or has embodied the body is to be granted the power of control over his or her body parts. Indeed, in current Dutch debates on such issues as organ donation and transplantation the question of ownership is seldom raised. Some of the authors have suggested simply to disregard the issue of ownership and to assume the embodied person has the power of control over his own body, allowing him to donate his blood and other organs under specific conditions ([3], p. 193).

Indeed, in the most recent laws as well as bills pertaining to the human body, its organs and its tissues, the term of 'property' is not mentioned. For example, in the new Law on the Disposal of the Dead, the term 'ownership' is absent. The same is true for the 1991 Advisory Report of the National Council for Public Health on Standardization of Research with Human Tissues. In debates on these new bills the issue of ownership is (intentionally) disregarded and the discussion is focussed completely on the limits of the right to self-determination: Is the removal of organs from a dead body allowed only after requiring an informed consent by the donor that was communicated by means of a donor card? May consent be presumed? Should living donors be allowed to donate their organs when doing so would endanger their own health? May a person sell his blood? In sum, which decisions about one's own dead body and parts shall be deemed socially and legally unacceptable?

We conclude that in Dutch law the concept of ownership of the human body plays no significant role. Because of theoretical inconsistencies and pragmatic difficulties in application, the body=property theory has few advocates in The Netherlands. However, bioscientific and biotechnological progress and the use of body parts in research and therapy raise enormous problems for the legal system, particularly if the body=property theory is rejected.

Creighton University
Omaha, United States
Catholic University of Nijmegen,
Nijmegen, the Netherlands

BIBLIOGRAPHY

1. Akkermans, P.W.C. (red.): 1987, *De Grondwet. Een artikelsgewijs commentaar*. Tjeenk Willink, Zwolle.
2. Bemmelen, J.M. van and Hattum, W.F.C. van: 1954, *Hand- en leerboek van het Nederlandse strafrecht*. Gouda Quint, Arnhem.

3. Frankena, H.C. and Graaf, F. de: 1990, 'Grondrechten en eigendom van bloed. Het NCAB-rapport over grootschalig AIDS-onderzoek', *Tijdschrift voor Gezondheidsrecht* **14**, 185-193.
4. Herten, J.H.S. van: 1984, 'De rechtspositie van lichaam, lijk, stoffelijke resten, organen en niet-menselijke implantaten', *Weekblad voor Privaatrecht, Notariaat en Registratie* **5689**, 155-164.
5. Hoge Raad: 1989, HR 8/11/88, *Nederlandse Jurisprudentie* NJ 1989, 667.
6. Hoge Raad: 1990, HR 2 juli 1990, *Nederlandse Jurisprudentie* NJ 1990, 751.
7. Leenen, H.J.J.: 1978, 'Recht op eigen lichaam', *Tijdschrift voor Gezondheidsrecht* **2**, 1-8.
8. Leenen, H.J.J.: 1988, *Handboek Gezondheidsrecht*. Samson, Alphen a/d Rijn.
9. Nationale Commissie AIDS Bestrijding: 1989, *Advies over grootschalig HIV-seroprevalentie onderzoek op anonieme basis*. NCAB, Amsterdam.
10. Noyon, J., Langemijer, G. and Remmelink, J.: *1992*, *Het Wetboek van Strafrecht*. Gouda Quint, Arnhem.
11. Oltshoorn-Heim, E.T.M.: 1990, 'Lichaamsmateriaal, epidemiologisch onderzoek en toestemming: de discussie, de argumenten, een conclusie', *Tijdschrift voor Gezondheidsrecht* **14**, 174-184.
12. Oven, J.C. van: 1926, 'Het recht op eigen lichaam', *Nederlands Juristenblad* **2**, 17-23.
13. Petit, Ch.: 1950, 'Lichaam en lijk als voorwerpen van rechtsbetrekking', *Themis* **12**, 428-435.
14. Praag, A. van: 1932, 'De rechtspositie van het stoffelijk overschot', *RM Themis* **93**, 88-98.
15. Rb Haarlem: 1987, Rb Haarlem 5-12-1986, *Nederlandse Jurisprudentie* NJ 1987, 549.
16. Rb Maastricht: 1990, Rb 4 & 20-10-1989, *Nederlandse Jurisprudentie* NJ 1989, 914 & NJ 1990, 227.
17. Reijntjes, J.M.: 1991, 'Rondom DNA. Over opsporing en bewijsvoering', *RM Themis* **6**, 267-284.
18. Rombach, J.: 1963, 'Prae- en postpersoonlijkheidsrechten en -plichten', *Weekblad voor Privaatrecht, Notaris-ambt en Registratie* **4774, 4775, 4776**, 297, 313, 329-299, 316, 331.
19. Tak, P.J.P.: 1990, *DNA en strafproces; een rechtsvergelijkend onderzoek naar de grenzen van het onderzoek aan en in het lichaam*. Gouda Quint BV, Arnhem.
20. Tweede Kamer: 1978-1979, 'Nota over de vraag of een bepaling over het recht op onaantastbaarheid van het menselijke lichaam in de Grondwet zou moeten worden opgenomen', *Handelingen Tweede Kamer* **Bijl. 1978-79, 15463**, nr. 2.
21. Tweede Kamer: 1989-1990, 'Aanhangsel', *Handelingen Tweede Kamer* **1989-90, 27/10/89**, 173-174.
22. Wery, C.F.: 1976, 'Beschikken over eigen lichaam en lichaamsdelen', *Ars Aequi* **25**, 305-315.

ANNE FAGOT-LARGEAULT

OWNERSHIP OF THE HUMAN BODY: JUDICIAL AND LEGISLATIVE RESPONSES IN FRANCE

I. INTRODUCTION

This essay is meant to be informative more than argumentative. The author does not express her own opinions but tries to sketch the so-called 'French law and ethics' of the identity relation between the person and his/her body, implying the principles of immunity or inviolability ('immunité') and unavailability or indisposability ('indisponibilité') of the human body, with the consequence that therapeutic or other uses of human blood or body parts should be non-profit. Of course, dissenting opinions have been expressed in France, but on the whole a broad consensus seems to exist, and it is expected that Parliament will pass further 'legislation on bioethics' systematizing the doctrine in law [34,35,36].

The present essay concentrates on the philosophical, ethical, and legal issues concerning the status of the human body, in relation to human rights. Very minimal information is given about the resection, handling, storage, or donation of human tissues or body parts, about blood transfusion, or about organ transplantation.

The essay is divided into five sections:

- overview of the French legal situation prior to 1988;
- experimenting on human subjects (law of 1988);
- the ethical status of the human body (recommendations by the National Consultative Ethics Committee, around 1990);
- 'from ethics to law': towards a legal status of the human body, a series of official reports; and
- legislation on bioethics (1994).

Only the period anterior to the 1994 legislation is covered here. The transition 'from ethics to law' (1989-1993) and the contents of the 1994 legislation are analyzed elsewhere [see, 19].

H.A.M.J. ten Have and J.V.M. Welie (eds.), Ownership of the Human Body, 115–140.

II. OVERVIEW OF THE FRENCH LEGAL SITUATION PRIOR TO 1988

Human Rights in General

The French *Constitution* of the 5th Republic (Oct. 4, 1958) in its Preamble states the attachment of the French people to human rights, as defined by the Declaration of Human Rights of 1789 and by the Preamble of the Constitution of the 4th Republic (1946). All three documents currently have full constitutional value, as reasserted by the Constitutional Council on July 16, 1971.

The French revolution abolished serfdom (Aug. 4, 1789) and slavery (Feb. 4, 1794). Liberty as a human right is interpreted as implying that no human being may be sold or bought, and that no human being may sell his/her own person. The Declaration of Human Rights included in the *Constitution* of 1793 explicitly stated that no person can be alienable property:

> Art. 23 – Tout homme peut engager ses services, son temps. Mais il ne peut se vendre ni être vendu. Sa personne n'est pas une propriété aliénable (Constitution, 1793).

The *Civil Code* [3] until 1994 did not explicitly give a status to the human body (or to human embryos); it separated persons from property, and stated that only property may be the object of an agreement. Agreements are contracted between persons; the object of the agreement may not be any person. For example, an agreement between a couple and a surrogate mother, the object of which would be that the surrogate mother will be inseminated, carry the pregnancy, deliver the baby, and give (or sell) it to the couple is void, because neither a woman's uterus, nor a baby, can be the object of an agreement (Civ. 1ère, 13 déc. 89, D. 90). Another example: an agreement about the painting of a tattoo on a person's skin is void (Trib. gr. inst. Paris, 3 jun 69, D. 70). In other words: the human body is not a commercial object.

> Art. 1128 – Il n'y a que les choses qui sont dans le commerce qui puissent être l'objet des conventions (Code civil).

The *Penal Code* [5] separates 'Crime and offence against persons' (Title II, Chap. 1) from 'Crime and offence against property' (Title II, Chap. 2). The Chapter on 'Crimes and offences against persons' (murder, infanticide, poisoning, injury, kidnapping, forgery, etc.) includes the 1946 law on the prohibition of procuring sex and brothels, the 1965 law prohibiting the use of stimulants in athletic competitions, the 1967 law on birth control, the 1975 law on termination of pregnancy, and the 1978 law on the protection of private data.

The *Code of Work* [4] states that an agreement between a worker and an employer may only be for a specific enterprise and/or for a definite amount of time; the object of the agreement is the worker's services, not the worker himself (his/her person).

> Art. L. 121-4 – On ne peut engager ses services qu'à temps ou pour une entreprise déterminée (Code du travail).

The *Code of Public Health* [6] is composed of nine books. I shall only give here brief comments on books 1 and 3.

Book 1 on the 'general protection of public health' includes the list of compulsory vaccinations, and other preventive measures against epidemics. In spite of registration of children in elementary school being conditional on proper vaccination, and compensation for possible prejudice caused by compulsory vaccination having been made automatic (Loi n 64-643 du 1 Juil 1964), there has long been in France a minority of libertarian parents boycotting compulsory vaccination and managing to circumvent the law. Even among the medical profession there is a minority with ambiguous feelings towards compulsory reporting of cases and mandatory vaccination. However, arguments against compulsory measures against transmitted diseases rely more often on the right to privacy than on the right to risk one's body.

Book 3 on the prevention of 'social plagues' ('fléaux sociaux'), includes Titles ('Titres') on tuberculosis, venereal diseases, cancer, mental disorders, alcoholism, drug abuse (since 1970), HIV infection (since 1983) and tobacco abuse (since 1991). The reporting of cases of drug addiction, possibly followed by a Court injunction resulting in coercive treatment, is viewed with skepticism by most health professionals. However, it is not generally argued that people may do what they please with their bodies, for example, getting intoxicated if they wish to do so. It is more often argued that coercive treatment is useless, anyway: after coercive treatment intoxication will recur.

Women's Rights in Particular, and Sexual Rights

In order to characterize more precisely what sort of rights French citizens have over their bodies, we shall here briefly envisage the issues of birth control, abortion, sexual abuse, prostitution, and transsexualism.

Up to 1967 *birth control* [26] was illegal in France. In 1967, a bill [26] authorized the production and distribution of contraceptive drugs. This bill was not implemented until Mrs Simone Veil, new Minister of Health, signed the proper statutory regulations in 1974.

Note that while the pill and the IUD are now legal in France, surgical means of contraception are not, unless prescribed by a medical doctor and justified by the person's medical condition. Vasectomy, even performed at the request and with the full consent of a male, is deemed to be a mutilation, and punishable as such (five to ten years in prison: Art. 310 of the Penal Code). Ligation of the tubes, being easier to justify by medical arguments connected with the female's health, is less rarely performed. Clearly French citizens do not have a right to surgical contraception as such: here there is a distinct limit to the right they have over their bodies.

Up to 1975 *termination of pregnancy* was illegal and considered a crime. A law voted in 1975 [28] 'decriminalized' abortion, while stating that respect is due to human beings 'from the very beginning of life':

> Art. 1 – La loi garantit le respect de tout être humain dès le commencement de la vie. Il ne saurait être porté atteinte à ce principe qu'en cas de nécessité et selon les conditions définies par la présente loi.

Section I of the 1975 law states the conditions under which voluntary termination of pregnancy (IVG) is permissible before the end of the tenth week of fetal development. The intervention must be performed by a medical doctor, after the woman has been fully informed of the risks and of her rights in case she decided to keep the child, and after she has consulted with a social worker. The woman's written consent is accepted only after a delay of one week. Women under 18 years of age need the consent of their legal guardian.

Section II states the conditions under which termination of pregnancy may be performed at any time during pregnancy for a 'therapeutic motive' (ITG). The therapeutic necessity (threat on the health of the mother, or high probability that the child will suffer from a serious and incurable disease) must be certified by two trained medical doctors, one of whom must be registered on a list of 'experts' by a Court of justice.

Any propaganda in favor of abortion is prohibited. Centers for family planning and birth control counseling must be non-profit. There is no mention in the 1975 law of women being free to make decisions about their bodies.

What about *sexual rights*? Up to 1946 brothels were legal. Prostitutes had to register with the police. In 1946, brothels were officially closed. Prostitutes ceased to have the obligation to register with the police. Procuring, promoting or encouraging prostitution, keeping a hotel in which prostitution occurs, became more heavily punishable. This was confirmed by Parliament in 1975 [23]. In brief: currently in France prostitution is not illegal, but profiting is.

A piece of legislation voted in 1980 [32] defines indecent assault and rape. Rape is a crime, punishable by five to ten years in jail, or more, in case there are aggravating circumstances, such as the abuse of a child (or any vulnerable person) by an individual (e.g., a parent) having authority on him or her. Only in recent years have a few cases of sexual abuse actually been prosecuted.

Does one have the right to change sex? The question of transsexualism has been much debated in recent years ([20,21,1,2]). There has been conflicting jurisprudence. In practice, since around 1978, some medical and surgical teams have cautiously accepted the responsibility of taking transsexuals into care, and to induce hormonal and morphological modification. Some Courts have then allowed transsexuals to reconcile their legal status with their physiological and psychological status, after irreversible surgery had been performed. But the jurisprudential situation is rather dissuasive, and legislation is now deemed necessary.

Body Parts, and Images of the Body

A bill passed in 1970 [27], which treats mainly of detention, remand in custody, judicial power, also states in its Part 3 entitled 'Protection of privacy' a new disposition to be included in the Civil Code:

> Art. 22 – [...] Chacun a droit au respect de sa vie privée.

That is deemed to imply, among other things, that my voice (when I speak in a private setting) or my image (when pictures of films are taken in a private setting) may be used only with my explicit consent.

The notion of explicit consent comes again in the law [30] of 1978 with a view to protect citizens against possible detrimental consequences of the filing and keeping computerized records of personal data. The law specifies the conditions under which a file or data base including personal data may be authorized. Any file must be declared and is controlled (and may be investigated) by a national commission called 'Commission Nationale de l'Informatique et des Libertés' (CNIL). No sensitive data may be recorded on a person, unless the person has explicitly consented to the recording. The law states that any person having properly identified him/herself has a right of access to any personal information kept on him/her, a right to have it corrected or deleted, a right to obtain a free copy proving that the correction has been made. There is however one major exception: citizens have only indirect access (through a medical doctor) to medical informations about themselves. In fact, confidential medical records (covered by professional secrecy) are the property of the institution (hospital) or of the medical officer, that is, the property of whoever keeps the record, not of the patient, and they are not supposed to be communicated directly to the patient. Such a proviso is meant to be protective. There is a minority of patients who claim that it is offensive, and that access to all personal medical data should be direct rather than indirect.

It has long been permitted (and encouraged) in France to dispose of one's dead body and donate it posthumously for 'scientific use' (in fact, for the training of medical students). The first law which organized the donation of organs to be grafted on live receivers was the 1949 law [24] permitting the removal of corneas from dead bodies on the condition that the deceased had explicitly stated in his will that he donated his corneas. He was not allowed to donate to a specified person, only to the community, that is, to a public or private institution promoting transplantation.

At the beginning of the 20th century, the transfusion of blood was performed 'from arm to arm', and direct payment of the donor by the receiver was quite common. With the development of blood preservation techniques, people became accustomed towards the middle of the century to donate their blood to the community, through an institution which processed, stored, and dispatched it as needeed; the payment of donors was still common rule. In 1949 (following world war II) a Federation of French blood donors announced that they were ready to donate their blood, rather than "selling it for money or for rationing tickets" ([38], p. 34). The law of 1952 which organized the 'therapeutic use of human blood' ruled out a free-market situation, and stated that the entire process of preparation and distribution of blood products should be non-profit. The Ministry of Health was in charge of setting prices. The law was always interpreted as implying that prices should take into account only the work invested in the processing, not the human material. Therefore blood had to be donated and donors could no longer be paid.

The well known law of 1976 [29] on the removal of organs for transplantation systematized the doctrine and applied it to all bodily organs. Only three of its main dispositions will be mentioned here. The removal of organs for transplantation from living donors is authorized (i.e., mutilation of a human body is permitted), provided the donors are adults without a mental handicap, and they give a 'free and explicit' consent; restrictive conditions are imposed on organ donation from minors (Art. 1). Consent is presumed for the removal of organs for transplantation from dead bodies, unless the dead person had made it known while he/she was alive that he/she refused (Art. 2). Organ donation is free, it may not involve any payment. The donor may be compensated for, say, his travel costs if he has to travel in order to donate, but the organ itself may not be paid for (Art. 3).

To summarize: although there are a few exceptions (e.g., for women's milk, or for the donation of a kidney to a sibling), the situation in the eighties is that French citizens are in general allowed to donate parts of their bodies (kidney, bone marrow, gametes), but they are allowed neither to be remunerated for it, nor to choose the beneficiary of the gift (they must donate to the community, via a medical institution; as a rule the gift must remain anonymous). Only the 1949 law uses the expression 'volunteer donors'. On account of its presuming consent, the 1976 Caillavet law has been interpreted by some as meaning that the (at least, dead) bodies of French citizens are the property of the community, or of the State; but such an interpretation is somewhat forced because in all cases the law stipulates a right of refusal.

III. EXPERIMENTATION ON HUMAN SUBJECTS

The Ethics of Experimentation on Human Subjects

Untill 1988, experimenting on human subjects for scientific purposes had no specific legal status in France, which meant that casualties due to research procedures would fall into the category of voluntary injury, and be liable to penal pursuit. Some lawyers had even deemed any kind of human experimentation unconstitutional, on the ground that the French Constitution guarantees the protection of health. In actuality, casualties due to biomedical research were covered up and turned into 'accidents' of therapeutics. That did not mean researchers were dishonest, for a clear awareness of acts of research being distinct from acts of therapeutics and requiring a particular ethics was rare, even within the medical profession.

In the early '80s, however, local and/or institutional research ethics committees mushroomed here and there, especially in French university hospitals. One of their concerns was to protect medical patients from being used for research purposes, and possibly injured, without knowing to what extent research procedures had been necessary from a therapeutic point of view. When in 1983 the National Consultative Ethics Committee for the Life and Health Sciences (CCNE) was established by President F. Mitterrand, it issued a report and recommendation on ethical problems posed by therapeutic trials on human beings ([9] n 2, Oct. 1984). This 1983 recommendation stated that there is a moral obligation for the medical profession to try new treatments ('devoir d'essai') according to the best standards of scientific methodology (including randomization of subjects, blind assessment of results, etc.), but that it could never be done without the free and informed consent of patients recruited for the trial. The same recommendation said that patients could only be used for research pertaining to the treatment of the disease from which they suffer. It mentioned the possible use of 'healthy volunteers' in medical trials, and declared that a law was needed to specify under what conditions that would be permitted: among other things, since healthy volunteers were healthy, and their health had to be protected, they could only be used in research involving 'minimal risk'. On the other hand, healthy volunteers should sign a contract with the researchers ('contrat d'essai'), and should be compensated (but not remunerated) for their contribution.

The report explicitly referred to the New York International Pact on civil rights to which France had adhered in 1976, to the Nuremberg Code (1947), to the Helsinki (1964) and Tokyo (1975) declarations of the World Medical Association, and to the recommendation issued by the World Health Organization in Manilla (1981). It made it clear that the protection of research subjects was an aspect of the protection of human rights.

In 1985, A. Milhaud in Amiens performed an experiment on a patient in a state of chronic vegetative coma, and claimed that the systematic use of such patients for medical research should be 'authorized by the CCNE' on account of the fact that such patients were 'virtually perfect human models, actually intermediate between animal and man'. The CCNE retorted with a report and recommendation ([9] n 7, Fev 1986) strongly disapproving of Milhaud's experiment, and stressing that vegetative patients are fragile human beings in need of special protection, who could in no way be 'treated as a means towards scientific progress, whatever the interest or value of an experiment which does not aim at improving their state of health' (Avis, par. 2).

In 1987, the CCNE was consulted by the Spatial Centre in Toulouse on a project (being a part of a conjoined CNES-NASA larger study) of experimenting on healthy volunteers to determine the physiological changes occurring in subjects maintained for six weeks under conditions of microgravity. Interestingly, the CCNE declared ([9] n 11, Dec 1987) not only that the project was scientifically sound and interesting, and could be carried out, but that the (carefully selected) volunteers should get paid a fair salary, their active participation being more comparable to 'work', than to the 'trade of their bodies'.

The statement that being a research subject may be compared to being a research worker, to the extent that one is an 'active participant', and not merely a passive object of research, recurred in a longer document on ethics and knowledge [8] in which the CCNE investigated the modes and manners, the funding and settings (sports, military, etc.), the philosophical groundings and the ethical principles and limits of biological research on human beings, aimed at acquiring new knowledge. This document proposed that human beings subjected to investigational procedures be considered as 'research partners'. But, although it admitted of the possibility of work contracts between research subjects and investigators, it fell back on the idea that the absence of remuneration (not of compensation) is in general a condition for human subjects to be treated, and treat themselves, with respect. In the meantime, paying a stipend to subjects of biomedical research had been ruled out by law.

The 1988 Legislation on the Protection of Persons Undergoing Biomedical Research

In December 1988, the French Parliament passed a law 'on the protection of persons undergoing biomedical research' [33], which legitimated research on human beings for scientific purposes in the field of biology and medicine, and stated the conditions under which it would be permitted. Regional committees in charge of scrutinizing the conformity of research projects with legal dispositions ('Comités Consultatifs pour la Protection des Personnes dans la Recherche Biomédicale': CCPPRB) have functioned since 1991.

The preamble of the 1988 law states that:

> Clinical trials or experiments organized and conducted in man with the objective of developing medical or biological knowledge are hereby authorized ... (L. 209-1).

It then distinguishes between two kinds of research:

> Biomedical research of which a direct benefit is expected for the person undergoing such research is hereby to be known as biomedical research of direct benefit to the individual. All other forms of research involving either sick or healthy persons shall be described as without direct benefit to the individual (L. 209-1).

The law very sensibly admits of non-therapeutic or 'without direct benefit' research conducted on medical patients. This was a shock for many citizens, and especially many lawyers. In spite of restrictive conditions specified in the law, ever since the law was adopted there was pressure put on Parliament to back up and to prohibit any research 'without direct benefit' to patients, particularly emergency care patients.

Part I of the law states general conditions under which bio-medical research may be conducted on human subjects. Preliminary conditions are that the project (1) should be 'based on the latest scientific knowledge and on sufficient preclinical testing', (2) should involve no risk 'out of proportion' with the expected benefit for subjects or with the interest of expected results, (3) should contribute to further scientific knowledge (L. 209-2). Actual conditions are that the research should be conducted (1) 'under the direction and supervision' of an experienced physician, (2) within an adequate technical setting and with proper scientific rigour and safety (L. 209-3). Restrictive conditions are put on the recruitment of three categories of vulnerable subjects: (1) no research 'without direct benefit' on pregnant or nursing women unless it 'carries no foreseeable risk' and is 'of value to scientific knowledge on pregnancy or lactation'; (2) no research on 'persons deprived of their freedom by a judicial or administrative decision' unless 'a direct and major benefit to their personal health is expected'; (3) protection from unjustified research 'without direct benefit' for persons whose autonomy is questionable, such as children or hospital patients. This third case being the contested case, let me quote literally:

> Minors, adults under guardianship, persons in medical or social establishments and patients in emergency situations may be solicited for bio-medical research only where a direct benefit to their health is expected. However, research which does not have direct benefit to the individual is permitted if the following three conditions are satisfied: – where such research presents no foreseeable risk to their health, – if it is of value to persons possessing the same age, illness or handicap characteristics; – and cannot be conducted otherwise (L. 209-6).

Article L. 209-8 states the general compensation principle:

> Biomedical research does not generate any direct or indirect financial gain for the persons undergoing it over and above the reimbursement of expenses incurred.

Part II of the law states that there shall be no bio-medical research on any human subject without the explicit consent of the subject (or of his parents or guardian if the subject is under guardianship). Consent should be given 'in writing' (or, in case that is impossible, witnessed by an independent party), after the subject has been fully informed. Provision is made for the physician withholding 'information related to the diagnosis', in 'exceptional cases where in the sick person's own interest the diagnosis of his illness has not been revealed to him'; but the possibility that such information be reserved has to be specified in the research protocol, so that the research ethics committee will know about it and may object to it. Here again the contested case is the case of research conducted on emergency patients, because the law allows for the possibility first to start the research, and inform them later, provided the ethics committee has approved such a protocol.

Part III of the law includes provisions for the review of research protocols and for the enforcement of reviewing procedures. Article L. 209-11 institutes review boards (CCPPRBs).

Apart from the fact that CCPPRBs are strictly consultative, they are somewhat similar to courts' juries. Their final composition is determined by lot. Deliberations are not public. Committee members must 'maintain strict secrecy over the information to which they may have access as a result of their function' (L. 209-11). They don't get paid for their function. They get reimbursed for justified commuting expenses, and compensated (modestly) for reporting on specific protocols.

CCPPRBs may be said to exercise a semi-democratic control on the way human beings are used in scientific (biomedical) research:

> The committee delivers its opinion on the conditions necessary to ensure the validity of the research, notably the protection of the persons undergoing it, their information, how their consent is to be received, any sum to be paid to them, the general relevance of the project, and the appropriateness of the means used, in view of the objective sought, as well as the qualification of the investigator(s) (L. 209-12).

The Ministry of Health is to be informed of any negative opinion on a project, and may 'suspend or prohibit' the carrying out of a project (L. 209-12).

Part IV of the law is entirely devoted to specifying the conditions under which research 'without a direct benefit' for subjects may be undertaken. Such research "should not carry any serious previsible risk for the health of persons undergoing it" (L. 209-14). Potential candidates are submitted to a preliminary medical examination. Requiring that they be covered by the National health insurance amounts to excluding the long-term unemployed or clandestine immigrants from recruitment and exploitation. Subjects may be compensated for their participation, unless they are "minors, adults under guardianship or persons in medical or social institutions" (L. 209-15), in which cases any compensation is prohibited. Volunteers are not allowed to participate in more than one trial at a given time, after the trial is over there is an 'exclusion period' during which they may not register in another trial (L. 209-17), and the total sum they are allowed to receive for compensation during a given year is "limited to a maximum set by the Minister of Health" (L. 209-15). The enforcement of such provisions obviously entails a strict control, and the national filing of volunteers in a data bank, for a period of one year from the time of enrolment in a trial. (The CNIL has controlled the acceptability of the filing procedure, and approved the filing after some amendments: [11], Délibération n 90-85, dated 26 Jun 90.)

Part V of the law makes provision for penal sanctions (imprisonment and/or fine) in four cases: violation of clauses pertaining to subjects' consent, violation of clauses relative to volunteers not being allowed to cumulate trials, research carried out without the protocol having been submitted to a CCPPRB, trial pursued after it has been suspended or prohibited by the Ministry of Health.

Worries and Queries about the 1988 Law

To the extent that the 1988 law was meant for the protection of subjects of biomedical research, it imposed stringent constraints on research sponsors and investigators, and could be perceived by professionals as a menace on the feasibility of any human research, and therefore, on the possibility to acquire any effective knowledge on the functioning of the human body. To the extent that the 1988 law made human experimentation permissible and gave a positive appreciation of biomedical research, it could be perceived as a menace on the personal rights and safety of citizens, particularly of the sick and vulnerable ones. Thus the law was much contested and criticized by all parties.

Medical professionals and researchers were unhappy because: (1) they claimed that the new legal dispositions were so complicated that research would be paralyzed, (2) they did not like the idea of asking the subject's (written) consent, (3) they got scared at the perspective of possible civil or penal pursuits being engaged by (informed) subjects, as a consequence of research procedures resulting in accidents or incidents. Patients should 'trust' their doctors, and doctors could'nt strive towards anything else than the 'good' of their patients. In some specialties (such as oncology) lobbies were formed and claimed that seeking the patient's consent was impossible, preposterous, or a hazard for the patient's health. Pediatric oncology simply wanted to be exempted from the obligation to comply with the rules. Anxiety apparently calmed down as people got used to the new state of affairs.

Patients' associations criticized the 1988 law on at least three grounds. First, the law gives way to medical paternalism, through allowing doctors to judge what kind of crucial information on their diagnosis should be concealed from patients. Second, the law legitimates a kind of exploitation of vulnerable persons through permitting (even though under restrictive conditions) that medical research 'without direct benefit' be carried on categories such as: children, adults under guardianship, emergency patients, the aged in retirement homes, and in general persons who stay in hospitals and are dependent on medical care. Third, an aggravating factor is that when such dependent persons are used for research, they cannot even be compensated. It was argued that such provisions were an insult to patients' dignity, and that granting the status of human guinea pigs to incompetent patients was in total contradiction with the Nuremberg code of 1947.

Harsh objections came from the side of lawyers. There appeared to be a discrepancy between the French Civil Code excluding the possibility of an agreement the object of which is the body of some person, and the new law instituting a weird kind of contract signed between subjects and investigators, the object of which is the carrying out of a research procedure on the subject's body (contract which stipulates that the subject may withdraw from the research at any time without incurring any liability!). Lawyers also essentially questioned the possibility left open by the law to start 'cognitive' investigations in emergency situations without asking anyone's consent. Some suggested that the new legislation served the purpose of organizing human experimentation better than the purpose of protecting human rights (see: [39]).

Especially interesting for our topic were the (somewhat jesuitical) discussions around the notion of compensation. That lending one's body to scientific research for pay is similar to prostituting one's body, is a comparison latent in traditional French culture. In the eyes of most commentators, if allowing people to be used for medical research was objectionable (particularly when they cannot give full consent), allowing people to be used for research and be paid for it was worse. It was like going back to slavery or legitimating another form of pornography. The human body could not possibly be the object of a commercial agreement. Indeed members of Parliament could not ignore the fact that, without some form of a payment there would be a shortage of volunteers in medical research. But that meant, payment added an incentive, for it induced people to accept what they should never accept for themselves, except out of altruism and devotion to the common good. Greed endangered freedom of the will. Hence the provisos: (1) vulnerable persons had to be protected against the temptation (they were not to get any payment), and (2) healthy volunteers were not to get any form of payment which would look like a 'benefit' (be it a good salary or a risk premium); they were to get exact 'compensation'. What for? The law says prudently: "for their participation" (L. 209-15). One common interpretation says: "for their time".

IV. THE ETHICAL STATUS OF THE HUMAN BODY

Reports and recommendations by the National Consultative Ethics Committee (CCNE)

Several documents issued by the CCNE [10] typically express a doctrine of the human body inherent in French culture at least from the French revolution of 1789. The first and most interesting of those is the document on 'biomedical research and respect for human persons' [7], which was prepared by a working group chaired by the philosopher, Lucien Sève. This document emerged from an effort made by members of the CCNE to conduct a reflection on their own principles, and bring out some of the (common?) foundations of their ethical views. As it remains a reference document for the CCNE itself, it is worth analyzing in some detail.

The document defines 'person', 'individual', 'human subject' (par. 2.2). It distinguishes 'personne de fait' (factual person) from 'personne de droit' (the subject in relation to other subjects). A human person has 'bodily' and 'other than corporeal' dimensions (par. 2.2.4). A person cannot be reduced to his/her biological dimensions (par. 2.2.5). Respect is due to the dignity of the person as a whole (par. 2.3). Human dignity is dignity above any price (par. 2.3.2) and cannot be measured in financial terms. 'Money reifies everything it buys and essentially equalizes things' (par. 2.3.2). The human body is in no way a thing and cannot become property. 'The simplest morality as well as our law exclude that the human body may be appropriated as a thing' (ibid.). It is equally immoral to recruite research subjects through the lure of gain, to sell or buy human body parts on the market, to hire out one's womb. The text gets emphatic: let me first quote the French, and then comment.

> La vie humaine ne doit pas être matière à appropriation privée. L'accepter si peu que ce soit constituerait un irrespect intolérable de la personne, une violation radicale de notre droit, une menace de pourrissement pour toute notre civilisation.
>
> Aussi bien notre pays a-t-il la chance, née non d'un hasard capricieux mais d'efforts admirables, qu'à partir du principe de la gratuité du sang, en heureuse rupture avec la pratique de longue date existante qui autorise la vente du lait maternel et des cheveux, se soit édifié un vaste système de dispositions, d'institutions et de valeurs en vertu duquel le corps humain dans toutes ses composantes, sans être

hors échange, est hors commerce. Il y a là un acquis national inestimable, souvent envié à l'étranger, auquel il serait désastreux de toucher mais indispensable d'ajouter les compléments pratiques comme les étais juridiques aujourd'hui nécessaires, et dont une opinion mieux instruite de son insigne valeur doit être appelée à se faire la dépositaire. Comment ne serions-nous pas vivement préoccupés par ce qui menace présentement de l'ébrécher, quand nous voyons sous tant de latitudes diverses à quelles extrémités peut conduire une vague montante de vénalité?

En ces matières, la logique de l'escalade est impitoyable. On commence par tolérer le négoce apparemment anodin de cellules, on finit par subir la plus infâme des aliénations de la personne. Sans être démentis, des journaux ont pu faire état ces derniers temps, ici, d'un trafic portant sur des dizaines de milliers de foetus dont beaucoup auraient servi à élaborer une arme biologique affreusement sophistiquée, là, de l'achat à vil prix ou du rapt d'enfants de familles misérables pour les rassembler en réserve d'organes que l'on vendrait à prix d'or. En vérité, si de tels faits sont bien exacts, n'ajoutent-ils pas à l'horreur biomédicale solennellement stigmatisée à Nuremberg il y a tout juste quarante ans? Il faut dresser une digue contre cette marchandisation de la personne, et il n'en est pas d'autre que le principe intangible selon lequel le corps humain est hors commerce – ce qui pose l'importante question des mesures propres à faire prévaloir dans les domaines concernés les exigences d'intérêt public sur les logiques de marché. Nous sommes convaincus que la dignité continuera de rimer avec la gratuité – cette dimension constitutive de l'acte moral (par. 2.3.2).

Most noticeable in the above is the slippery slope 'argument' ("logique de l'escalade"): you start with what looks like an innocuous and minor trade in human cells, but you end up with a major and abhorrent alienation and exploitation of persons, and the corrupting of society. The argument is used to reject any compromise solution between the market situation and what ethics and 'the public interest' require. The human body, whether in whole or in parts (organs), cannot be sold. Human dignity is synonymous with the refusal of any merchandizing of bodies/persons. Doing it 'for free' is a constitutive dimension of any moral act.

This doctrine is applied in several reports and recommendations. The 'Avis' on problems arising from the development of the technological use of human cells, or of products derived from human cells ([9] n 9, Feb 1987) declares (and the document above confirms: [7], par. 3.2.2) that when a sample of blood, or of any human tissue, is taken in the course of a therapeutic act, (1) the sample may be considered as donated by the patient to the medical team or hospital, (2) if the medical team uses part of it to develop a product (such as a diagnostic test) which could be commercialized, the price on the market should reflect only the work put in and in no way the material which in itself has no price, (3) the patient should have no right on possible financial benefits derived from the product being sold (supposedly, the activity should be non-profit, that is, possible benefits should be reinvested in bio-medical research, exclusively aimed at producing new knowledge, and finally better therapeutics, any other purpose being banned). That you may not sell your body implies that you may not sell any of your cells:

> Les cellules provenant du corps humain ne peuvent être considérées différemment de tout autre élément de ce corps. Il n'existe donc pas de raison spécifique de les exclure du principe de refus de commercialisation. Il en résulte que l'homme ne peut être autorisé à vendre ses propres cellules ([9] n 9, Feb 1987, Report).

A recommendation and report 'on the non-marketing of the human body' ("sur la non-commercialisation du corps humain": [9] n 21, Dec 1990) rehearses the steps through which the ethics of respect for the human body was elaborated. It states that its basic principles are final, that they will probably not need to be reexamined, and that they agree with the French legal tradition. Its essentials are: my body is not my property (it's me); I may not alienate it or make money from it; no part of any body may be bought or sold (that would be an insult to human dignity: things have a price, persons are endowed with dignity); society has a right, nay an obligation to protect people from alienating parts of their selves, and particularly to prevent the most vulnerable populations from being physically exploited (selling one's organs or blood cannot be a matter left to individual choice but must be prohibited); when products derived from the human body are marketed their price legitimately reflects the amount of work invested in manufacturing them; their price has got nothing to do with the organ or bodily substance which remains basically free; etc. Key passages are:

> ... Ni le corps humain, ni une partie du corps humain, ne peuvent être vendus ou achetés. [...] Dire que le corps humain est hors-commerce,

> ou encore hors-marché, c'est formuler deux propositions complémentaires: d'une part le corps de l'homme, ou l'un de ses éléments ne peuvent être l'objet d'un contrat, d'autre part il ne peut être négocié par quiconque. Ainsi un organe tel que le rein, ne saurait être vendu par celui d'où il provient, et, fût-il cédé gratuitement, être vendu par un tiers quelles que soient les incitations du receveur éventuel ou de son entourage. Ces incitations pourraient aller jusqu'à engendrer des chantages sur les personnes dépendantes comme par exemple les détenus ou toute minorité dominée. Il y va de la dignité de l'être humain de ne pas tirer finance de son amoindrissement physique même temporaire.
>
> On perçoit ainsi toutes les conséquences qui au cas de solution contraire, pourraient être déduites de la misère économique dans la partie la plus vulnérable de la population et chez les populations les plus vulnérables ([9] n 21, Dec 1990, Avis).

The same principles apply, according to the CCNE, also to human genetic information, as is confirmed by a subsequent 'Avis' ([9] n 25, Jun 1991), challenging the claim of some groups to charge a fee for access to human sequence data. The principles entail that no sequence of the human genome should be patented because the human genome is a 'common heritage of the human race', as reasserted in an 'Avis sur la non-commercialisation du génome humain' ('on the non-marketing of the human genome': [9] n 27, 1991).

For awhile the doctrine even resisted the shock following the disclosure in 1991 of the French blood transfusion affair. The CCNE, in 1991, adopted the view that accidents of infection (with contaminants such as the aids viruses) through blood transfusion had been due to breaches of the principles of the correct ethics of transfusion, that is, to the fact that the authorities of the transfusion system had attempted to make profit. (Note that a French Senate commission, on the contrary, reckoned that accidents had for the most part been due to authorities sticking to the traditional ethics of unpaid donors being above all suspicion, and of solidarity prevailing over safety: see [38]).

The CCNE report and recommendation 'on blood transfusion from the viewpoint of the [principle of] non-marketing of the human body' ("sur la transfusion sanguine au regard de la non-commercialisation du corps humain": [9] n 28, Dec 1991) reasserts the 'fundamental values' which had inspired the organization of the French transfusion system (no payment, no profit, consideration for donors on account of the active solidarity they show with the ill and needy). Indeed, the document says that sticking to the ethics of non-profit does not exempt one from worrying about the safety of blood products. But clearly, in the hierarchy of ethical principles, the imperative of safety is viewed as secondary and a mere technicality, while it appears as a primary imperative that the human blood, as a part of the human body, may not be the object of a commercial agreement or profit. The report goes as far as objecting to the European regulation (n 89-381, Jun 1989) requiring for blood products similar certification procedures and standards of safety as are required for drugs, on the argument that the European document calls the human blood a 'raw material' and assimilates blood products to 'drugs', that is, to things, which is not acceptable. What is at stake, the report concludes, is that any commercial system will induce an inequality and exploitation of the most disadvantaged groups of the population, while the French system of solidarity 'enhances societal links'.

Enhancing societal links thus comes out as a basic moral imperative, from which it follows that the social benefit derived from individuals exchanging blood through a common pool which does not discriminate between them justifies imposing on citizens the constraint (and risk) to donate and receive anonymously and non-commercially.

The philosophical (or religious?) background of such a doctrine, which views individual bodies as members or organs of the social body, has never been developed by the CCNE.

Dissenting Voices

In the Roman Catholic tradition, the church is viewed as the (mystical) body of Christ (in the protestant tradition there is no such mediation of an institutional body between man and God). In the French (and English) tradition of monarchy, the King incarnated the nation, i.e., there was a kind of mutual incorporation of the King in his subjects, and of the subjects in their King. Both traditions merged in France: the Gallican church and the Catholic King embodied the 'patria' (or 'crown'), that is, the spiritually and politically structured community, the 'domain' of which could not be 'alienated', and was defended by soldiers 'incorporated' into the army. This notion of an organic community transcending individuals seems to have been secularized, and resumed rather than reversed, by the French Republic[1]. What the 1789 revolution brought about was the guarantee that no member of the community may freely dispose of the body of any member (not even his/her own), which reinforced rather than cancelled the notion of a vertical dependence of citizens from the State (relations between the parts had to be mediated through the whole). Admittedly the metaphor of the political body of which individuals are members was already present in many ancient texts[2]. But it was then a metaphor, as opposed to the organic realism of the political body found in modern Europe (see: [22]). While the 'French ethics' of the majority (as expressed by the CCNE) is a mixture of human rights and political (theological?) organicism, minority views advocate either more individual rights, or a stronger solidarity. We shall briefly sketch an example of each kind.

A libertarian attitude may go with philosophical dualism, and/or with political liberalism. For example, an essay by the economist B. Lemmenicier, took the stand that the human body is a machine, the owner of which is the mind (or spirit) residing in it. The owner should be deemed free to dispose of his/her machine the way he/she wants. Let us not confuse ethics with law, pleaded the author. We need a common law ensuring mutual respect and tolerance for our various ethics. Deciding to what extent my body parts deserve 'respect' is a matter of my ethics, provided the law prohibits others to use my body in ways that I do not want. If my ethics is compatible with my buying a newborn child from a consenting woman, or with selling my kidney or a few cells, I should not be prevented by law from doing it. Claiming that the trade of human body parts is 'morally degrading', he concludes, is as ridiculous as claiming that an opera singer who sings an aria he dislikes 'prostitutes' his/her voice (see Lemmenicier, in [18]).

The philosopher François Dagognet, at the other end of the spectrum, has argued in favor of strong communitarian views, linked with philosophical monism and political solidarism in the spirit of Comte or Hegel. Presuming consent for the removal of organs or tissues from dead bodies, he says, is an hypocrisy: the organs of the dead belong to the community, the recycling of organs and tissues should be systematic. We have an obligation of solidarity. There are people agonizing in need of organs, their misery is an injunction to transplant, and any organs available should be requisitioned. The gift of blood should be made mandatory, a civil service. Let us not fear that the replacement of defective body parts might consolidate a mechanistic conception of organic life: physiologically grafted fragments are actively reappropriated by the receiving self, and more generally the community of bodies is of crucial importance for the single body. A body does not exist in isolation. A brain does not function if not stimulated. The human body is not a thing closed on itself, but is in some essential way relational, and the exchange of body parts made possible by the technology of transplantation is a symbol of the community as an interrelational body (see, [13,14,15,16,17]).

Actually such dissenting proposals were scarce and had little impact at the time. From 1985 to 1993 the mainstream ethics was methodically elaborated through a series of five major public reports: the Alnot *et al.* report on artificial procreation (1986), the Council of State (Conseil d'Etat) report explicitly entitled "from ethics to law" (1988), the Braibant project on life sciences and human rights (1989), the Lenoir report (1991), and finally the report of the French parliamentary Office for the assessment of scientific and technological choices (1992). Then in 1992 three projects of legislation inspired mainly from the Lenoir report were submitted to Parliament by the French government. The first project, prepared by the Ministry of Justice, pertained to the dignity of the human body, and was meant to modify the Civil Code. The second project, prepared by the Ministry of Social Affairs, pertained to the use and exchange of body parts and products and to medically assisted procreation, and was meant to modify the Code of Public Health. The third project, prepared by the Ministry of Research and Technology, pertained to the protection of confidential data used in research, and was meant to modify the law [30] on freedom and the filing of personal data. Three laws ([34,35,36]) were finally passed in July, 1994, after extensive discussion and several shuttles backwards and forwards between the Senate and the Chamber of Deputies. The Codes were modified accordingly. For want of space the nuances of philosophical doctrines implicit in the reports and laws are not scrutinised here (for an analysis of the conception of the human body inherent in the post-'94 French legislation see: [19]).

V. CONCLUSION

In the Huxley memorial Lecture he delivered in 1938 on the notion of 'person', the anthropologist Marcel Mauss suggested that in western countries the right to dispose freely of one's body has been a more aristocratic right, and a right which became democratized much more slowly, than the right to dispose freely of one's mind. As an example, he says, people in serfdom were denied mastery over their bodies by Germanic law long after the Church had granted them mastery over their souls. Only free people, as opposed to serfs, were deemed to own their bodies (see: [37], par. 4). The elitist character of the right to dispose of one's body is also patent in the fact that fighting duels was a priviledge long-lastingly claimed by noblemen.

Traditional aristocratic ethics also includes the idea that hiring one's working force amounts to hiring oneself, and that the financial link between an employer and his employee is a servile link. The nobleman does not work for money. He will not strive to 'earn his living'. There are people who still think that way. An eminent member of the French National Consultative Ethics Committee declared publicly in December, 1989, that 'doing some thing for money amounts to enslaving one self'. From that perspective the idea of a 'just salary' makes no sense. No salary can be just if the very notion of a salary is intrinsically immoral. And the notion of a salary is immoral for those who equate with enslavement the work contract instituting a proportionality between the amount of work done and the amount of money earned. The 'liberal' professions used to be those in which the provider of a service would 'give' and not count his time and expertise, and would get in return a 'honorarium', or 'fee', the amount of which was supposedly left to the free will of the beneficiary. The 'liberal' medical doctor used to conceive of one self as equally devoted to all his patients, and giving them all the care they needed, without any financial consideration. He would not expect a fee from the indigent, and would leave others free to appreciate how much they could pay him.

There is obviously something left in the 'French' ethics of that aristocratic ideology, from the viewpoint of which *human dignity* implies that you will not stoop to hiring or selling your self for money, because you have a sense of your life being worthy in itself – a unique present you received from destiny, out of proportion with anything money can measure or buy. The (absolute) value of the gift of life also entails that, as Auguste Comte said, "we were born laden with obligations towards society", from which no contractual arrangements can untie us (see: [12]). The 'liberal' doctor who did not admit paying his patients with a service proportional in quality or time to the amount of money they could pay, at the same time felt an endless obligation towards the community to give all his very best to all his patients. He could not have declined giving his services, no more than the nobleman could have declined fighting a duel: a matter of honor.

Whether such an ethics of human dignity can be maintained in an era of recycling of human body parts, and human genetic engineering, is an open question. What seems clear is that in the French legal tradition, selling a piece of one's body, that is, of oneself, is deemed so utterly degrading of the person that there is a broad consensus to presume that it is the legislator's responsibility to prohibit it, that is, to protect individuals from the eventuality of doing it. The other side of the same coin is that, as body parts or elements are exchanged, they are not exchanged directly between individuals: all exchanges are mediated and regulated through the community so that, in a sense, human body parts may be said to be common property of the community.

Clearly French citizens are not the owners of their bodies. A 'philosophical' reason given by lawyers to justify such a situation is that the human body is (identical with) the human person. In a democratic society, all human beings have the status of persons. Persons are subjects. They contract with each other and trade-off things. Things can be the object of an agreement and things can be bought and sold. Persons cannot be the object of any such agreement. Trading persons, or parts of persons, or elements of persons in the market place, would turn subjects into objects, that is, subvert the foundations of the social order. Preserving the freedom of subjects involves maintaining (so to speak) all parts and bits of subjects within the realm of persons, and not let them slide into the realm of things. That, however, seems to prohibit exchanges of human body parts. One way to make exchanges possible would be to declare that some parts of the body are 'detachable' from the person. French lawyers reject such a possibility, by citing a slippery slope argument. They prefer another option. Exchanges are made possible by the community acting as the actual owner of all body parts, with the consent of individual persons.

Department of Philosophy
University of Paris-I
Paris, France

NOTES

[1] This paragraph has benefited from remarks by Suzanne Rameix (personal communication).

[2] Rameix enumerates Plato's *Republic* (V, 435c), the *Politics* of Aristotle (I, 2, 1253a 20-29), Epictetes' *Dialogues* (II, 5, 24-27), and the well known apologue of Menenius Agrippa in: Livy's *Ab Urbe Condita Libri* (II, 32).

REFERENCES*

1. Alméras J.-P.: 1989, 'Le transsexualisme', *Le Concours médical* **111** (32), 2762-2763.
2. Alméras J.-P.: 1991, 'Les transsexuels et le droit', *Le Concours médical* **113** (7), 565.
3. *Code civil.* Dalloz, Paris.
4. *Code du travail.* Dalloz, Paris.
5. *Code pénal.* Dalloz, Paris.
6. *Codes de la santé publique, de la famille et de l'aide sociale.* Dalloz, Paris.
7. Comité Consultatif National d'Ethique (CCNE): 1987, *Recherche biomédicale et respect de la personne humaine.* DF, Paris.
8. Comité Consultatif National d'Ethique (CCNE): 1990, *Ethique et connaissance. Une réflexion sur l'éthique de la recherche biomédicale.* DF, Paris.
9. Comité Consultatif National d'Ethique pour les sciences de la vie et de la santé (CCNE): 1993, *Xe anniversaire. Les avis de 1983 à 1993*, INSERM, Paris (includes all 34 recommendations and reports issued by the CCNE between 1983 and 1993. The same documents may be found scattered in the yearly *Rapports* of the CCNE, and in the quarterly *Lettre d'information du CCNE*; both published by DF).
10. Comité consultatif national d'éthique, *Rapport.* DF, Paris, yearly (from 1984 on).
11. Commission nationale de l'informatique et des libertés, *Rapport d'activité.*, DF, Paris, yearly (from 1980 on).
12. Comte A.: 1852, *Catéchisme positiviste.* Chez l'auteur, Paris.
13. Dagognet F.: 1989, *Eloge de l'objet. Pour une philosophie de la marchandise.* Vrin, Paris.
14. Dagognet F.: 1990, *Corps réfléchis.* Editions Odile Jacob, Paris.
15. Dagognet F.: 1992, *Le cerveau citadelle*, coll. Les empêcheurs de penser en rond. Delagrange, Paris.
16. Dagognet F.: 1992, *Le corps multiple et un*, Paris, coll. Les empêcheurs de penser en rond. Delagrange, Paris.
17. Dagognet F.: 1993, *La peau découverte*, coll. Les empêcheurs de penser en rond. Synthelabo, Paris.
18. *Droits, Revue française de théorie juridique*, 1991 (1): n° 13 – "Biologie, personne et droit", special issue.
19. Fagot-Largeault A.: 1996, Problèmes d'éthique médicale posés par de nouvelles techniques thérapeutiques: greffes d'organes, de tissus et de cellules, in: P. Livet, éd., *L'Ethique à la croisée des savoirs.* Vrin, Paris.
20. Gobert M.: 1988, Le transsexualisme, fin ou commencement?, *La Semaine Juridique*, Ed. G, n° 47, I, 43: 3361.
21. Gobert M.: 1990, Le transsexualisme ou de la difficulté d'exister, *La Semaine Juridique*, Ed. G, n° 49, I, 45: 3475.

* 'JO' = *Journal Officiel de la République Française*, Direction des Journeaux officiels, 26, rue Desaix, 7015 Paris, France. 'DF' = La Documentation Francaise, 29 quai Voltaire, 75007 Paris, France; fax +33 1 40 15 72 30.

22. Kantorowicz E.H.: 1957, *The King's two Bodies. A Study on Mediaeval Political Theology*. Princeton University Press, Princeton.
23. Loi n° 46-685 du 13 avril 1946 tendant à la fermeture des maisons de tolérance et au renforcement de la lutte contre le proxénétisme, *JO*, 14 04 46: 3138-3139; modifiée, Loi n° 75-624 du 11 juillet 1975 modifiant et complétant certaines dispositions de droit pénal, *JO*, 13 07 75: 7219-7225.
24. Loi n° 49-890 du 7 juillet 1949 permettant la pratique de la greffe de cornée grâce à l'aide de donneurs d'yeux volontaires, *JO*, 08 07 49: 6702.
25. Loi n° 52-354 du 21 juillet 1952 sur l'utilisation thérapeutique du sang humain, de son plasma et de leurs dérivés, *JO*, 22 07 52: 7357-7358.
26. Loi n° 67-1176 du 28 décembre 1967 relative à la régulation des naissances et abrogeant les articles L. 648 et L. 649 du code de la santé publique ('loi Neuwirth'), *JO*, 29 12 67: 12861-12862.
27. Loi n° 70-643 du 17 juillet 1970 tendant à renforcer la garantie des droits individuels des citoyens, *JO*, 19 07 70: 6751-6761.
28. Loi n° 75-17 du 17 janvier 1975 sur l'interruption volontaire de la grossesse ('loi Veil'), *JO*, 18 01 75: 739-741; modifiée, Loi n° 79-1204 du 31 décembre 1979 relative à l'interruption volontaire de la grossesse, *JO*, 01 01 80: 3-4.
29. Loi n° 76-1181 du 22 décembre 1976 relative aux prélèvements d'organes ("loi Caillavet"), *JO*, 23 12 76: 7365.
30. Loi n° 78-17 du 6 janvier 1978 relative à l'informatique, aux fichiers et aux libertés, *JO*, 07 01 78: 227-231.
31. Loi n° 78-753 du 17 juillet 1978 portant diverses mesures d'amélioration des relations entre l'administration et le public et diverses dispositions d'ordre administratif, social et fiscal, *JO*, 18 07 78: 2851-2857.
32. Loi n° 80-1041 du 23 décembre 1980 relative à la répression du viol et de certains attentats aux moeurs, *JO*, 24 12 80: 3028-3029.
33. Loi n° 88-1138 du 20 Dec 1988 relative à la protection des personnes qui se prêtent à des recherches biomédicales ("loi Huriet"): *JO*, 22 12 88: 16032-16035; Modifiée: Loi n° 94-630 du 25 Jul 1994, *JO*, 26 07 94: 10747-10749. Complétant le *Code de la santé publique*, Livre II bis, Art L. 209-1 à L. 209-23. Published with English translation: Bulletin Officiel n° 90-4 bis, Direction des Journaux Officiels, address above. Published with statutory and regulative dispositions: Bulletin Officiel n° 91-12 bis and 91-13 bis, same address.
34. Loi n° 94-548 du 01 Jul 94 'relative au traitement de données nominatives ayant pour fin la recherche dans le domaine de la santé et modifiant la loi n° 78-17 du 6 janvier 1978 relative à l'informatique, aux fichiers et aux libertés', *JO*, 2 Jul 1994, 9559-60.
35. Loi n° 94-653 du 29 Jul 1994 'relative au respect du corps humain', *JO*, 30 Jul 1994, 11056-59.
36. Loi n° 94-654 du 29 Jul 1994 'relative au don et à l'utilisation des éléments et produits du corps humain, à l'assistance médicale à la procréation et au diagnostic prénatal', *JO*, 30 Jul 1994, 11060-68.
37. Mauss M.: 1938, Une catégorie de l'esprit humain: la notion de personne, celle de 'moi', *Journal of the Royal Anthropological Institute*, LXVIII (Huxley Memorial Lecture). Repr. in: *Sociologie et anthropologie* (1966, 3ème édition augmentée). PUF, Paris, 331-362.
38. Sourdille J. & Huriet C.: 1992, *Rapport de la commission d'enquête sur le système transfusionnel français en vue de son éventuelle réforme, créée en vertu d'une résolution*

adoptée par le Sénat le 17 décembre 1991, Sénat, Seconde session ordinaire de 1991-92, Document n° 406, Annexe au procès-verbal de la séance du 12 juin 1992, Paris.

39. Thouvenin D.: 1989, Commentaire législatif. La loi du 20 Déc 1988: loi visant à protéger les individus ou loi organisant les expérimentations sur l'homme?, *Actualités législatives Dalloz*. **10**: 89-104, **11**: 105-16, **12**: 117-128.

PART IV

OWNERSHIP OF THE BODY: THEORETICAL PERSPECTIVES

KEVIN W. WILDES, S.J.

LIBERTARIANISM AND OWNERSHIP OF THE HUMAN BODY

I. INTRODUCTION

Biomedicine in the clinic and in research has been shaped by the technological developments of the last forty years, and for the foreseeable future it will continue to be shaped by these developments. Technology has not only enabled medicine to do better what had been done before, but it has also reshaped the possibilities of health care and human life. As biomedical technologies realize new possibilities for medicine the limitations of human existence are also challenged. Biomedical technology opens the door to new possibilities in curing illness, prolonging life, and altering the most basic structures of human life itself. In doing so, these technologies create new choices in almost every aspect of human life from reproduction to the way we die.

In providing a wide range possibilities biomedicine also raises moral concerns about how one *ought* to act. Many of the technological developments in biomedicine, for example, raise questions about how we think the human body should be treated. The body is not only the locus of medical intervention but it is also the source of materials for such interventions. From blood transfusions to the use of fetal tissue and whole organs the human body has become an important resource for biomedical intervention. However, the ways in which bodies are used often evoke strong moral reactions. For example, many hold that the body has "special meaning" [13]. The implication, of course, is that there are certain ways in which the body cannot be viewed or treated. However, these intuitions and claims about the body are not shared by all. Indeed, people often have radically different intuitions about the proper use of the body and the conceptual question is how one is to understand and judge the competing intuitions in general secular terms.

The moral significance of such claims can only be understood within the context of a moral discourse. Outside such a particular shared discourse or moral narrative moral intuitions cannot be understood. At the

H.A.M.J. ten Have and J.V.M. Welie (eds.), Ownership of the Human Body, 143–157.

level of general, secular discourse there is a loss of meaning for moral terms since there is no coherent common moral language, or set of values, which is shared. Absent a communal context of values and language such special concerns and meanings about the body are, literally, vacuous in secular discourse.

The first section of this essay argues that the postmodern crisis of secular moral authority renders appeals to intuitions about the 'special meaning' of the human body at best confused and ambiguous, or, at worst, meaningless. In the second part I sketch out the implications of the postmodern condition for understanding the human body and health care policy. Moral discourse at this level relies on the authority of moral agents and the discourse is purely procedural and empty of content. In the final section I draw a sharp contrast with the emptiness of the general, secular level, by turning to a view of the body within a particular community.

Confronted by a pluralism of moralities, rather than a single moral narrative, the only defensible moral position, in the secular context, is the libertarian position. Since one cannot assume that others will share the same moral values, rankings, and premises in resolving moral controversies (since reason cannot discover a contentful moral vision which all rational agents are compelled to accept), then one will have to rely on the moral authority of individual moral agents. The libertarian conclusion is not a covert appeal to the value of liberty for such an appeal would beg the question. Rather, it is the only rational conclusion one can reach in a secular society of 'moral strangers'. If God is silent, and reasons cannot discover a contentful morality then one will have to rely on the consent of those involved in a dispute if one is to act with moral authority. Controversies surrounding the ownership and use of the human body (e.g., use and sale of blood products, participation in human experimentation) can only be resolved, on a secular level, by the agreement of the parties involved.[1]

II. THE POSTMODERN PREDICAMENT

The term 'postmodern' has had a wide currency in contemporary intellectual circles and one of the difficulties with the term stems from its diverse usage. As used in this essay it is concerned with the sociological, cultural, and philosophical implications of the term for moral philosophy.

The sociological dimension of postmodernism is the factual observation that there is a widespread diversity of moral opinions and sensibilities. Here the laboratory of bioethics is instructive in that one has only to recall the many controversies of bioethics to realize the plurality of moral visions and values in contemporary western culture (e.g., abortion, euthanasia, the allocation of health care resources). Biomedicine often takes on moral dimensions in that health, as a value, is set within the other values of a person's life. In most cases, however, medicine evokes more than simply a relationship between a physician and a patient. It usually is undertaken in a cooperative venture often involving both private resources and those administered by governments on behalf of society. Because an understanding of health is so tied to human values, and because biomedicine is practiced in a public, diverse context, biomedicine is often a crucible of moral pluralism and controversy.

This pluralism of moral values reflects a profound change in western culture. Western culture has gone from a culture, homogeneously Christian and often intolerant of other views, to a culture of moral pluralism which expects a diversity of moral judgments about issues. The sociological description of pluralism in moral values is an indication of the conceptual difficulties posed by moral pluralism [17].

The conceptual difficulty of postmodernism for ethics lies in justifying the initial choice for moral judgments. If one understands the 'secular' to be a morally neutral framework [8], then the choice of a particular morality faces the dilemma of the choice of a starting point. One might hope, as the West has, that a content can be discovered by reason; that is, without appeal to a particular religion or culture. This hope has continued as bioethics has deployed different theoretical justifications borrowed from the history of moral philosophy. There have been numerous attempts to develop contentful theories which justify moral judgments (e.g., [6,12,22,23,26]). Each theory however faces two initial difficulties. The first is to justify its foundational starting point. Each theory of normative morality must presuppose some structure of reason on which to base its appeal (e.g., consequences, duty, nature) [3]. However, it is not at all clear why one should choose one element of moral experience over others as the basis for the structure of one's moral reasoning. Indeed one of the criticisms of moral philosophy from the feminist point of view has been a criticism of the assumptions made about the nature of reason [4,11]. Resolving this dilemma, however, will not address the second issue which is that each theory needs a content. Without a content (some expression of moral commitments) the theory will remain universal but be unable to deal with concrete dilemmas. With content the theory becomes particularized. Unless the agents share the same ranking of values they will not be able to agree on the proper resolution of a moral controversy.

In order to make judgements about choices in actions or policy a moral theory will require some ranking of values. Without such moral commitments the theory remains vacuous. For example, the first principle of the natural law (do good and avoid evil) ([1], I-II, 94, a.2), needs content if it is to guide practical reason. One needs to specify a ranking of goods if a moral theory is to aid men and women in developing resolutions to moral dilemmas. Theories often build in ways in which goods are ranked (e.g., the mechanism of preference utilitarianism, the basic goods of human flourishing, the lexical ranking of principles). However, in a secular world there is no canonical ranking of values which is shared by, or binding upon all as the secular framework is morally neutral it will admit many different starting points. Indeed, there is an overabundance of answers to the foundational questions about moral reason.

Bioethics has tried to avoid this scylla and charybdis by appealing to middle level principles without foundation in any particular theory. The appeal to middle level principles, however, is fraught with problems. First, there is the difficulty in determining what any of the principles mean. One suspects, for example, that when Beauchamp and Childress claim to have agreed on the meaning of the four middle level principles one is witnessing a slight of hand. For a preference utilitarian, like Beauchamp, 'autonomy' will be understood as the liberty to achieve certain goods. For a deontologist, like Kant, autonomy is concerned not with the pursuit of heteronomous desires but with acting in accord with the demands of reason imposed by the moral law. Childress, even though he does not endorse a Kantian deontology, relies upon right making and wrong making criteria that are independent of consequences. So while both Beauchamp and Childress speak of 'autonomy', the term is a placeholder for many different meanings in radically different languages. In the fragmentation of moral language the same word, or moral term, can mean many different things. Terms like 'justice', for example, can mean very different things to a communitarian or a libertarian. The meaning of moral terms will depend on the context and assumptions which frame their deployment. Such moral language, detached from particular moral frameworks, becomes evacuated of content in that outside of particular frameworks there is no way to control the meaning of such terms. They can then take on so many meanings as to become babel.

Much of the effort to regulate and constrain the use of the human body is embedded in a moral language of 'sanctity', 'dignity', 'justice', 'integrity', and 'solidarity' that assumes particular moral commitments. The meanings of these terms can only be controlled and understood within a moral narrative. While people may use the same terms, the terms can often have radically different meanings.

A second difficulty with the middle level principles approach is that it is never clear how the principles should relate to one another. Each of the principles is conceived of as *prima facie* binding. If the principles are all of equal weight, it is not clear how we are to decide, in a conflict situation, which principle to follow. A final difficulty facing this approach is that there is no clear justification as to why this set of principles should be canonical. One might wish to include other principles, such as 'sanctity of life', 'equality', the 'principle of solidarity', or the principle of *Menschenwürde*, which have been omitted. Even other attempts to resolve moral dilemmas by appeals to principles may use other principles. Frankena, for example, using a similar approach, develops a different structure and list of principles [10].

To resolve these three dilemmas, the middle level principle model would have to be radically recast in the context of a theoretical account which would define the meaning, relations, and justification of the principles. In Western philosophy principles have traditionally been a part of a comprehensive structure which begins with some first principle(s) and moves to secondary (middle level) principles ([1], I-II, 94, aa. 1-6). Seeking to avoid the difficulties of theoretical accounts, Beauchamp and Childress have attempted to excise the secondary principles from any type of comprehensive structure. They hope in this way to avoid the challenge of providing foundations. However, shorn of theoretical and contextual moorings the principles become incomprehensible and incoherent in resolving moral dilemmas.

A second attempt to offset the difficulties of moral theory has been the effort to revive moral casuistry [15]. Jonsen and Toulmin offer an historical account of casuistry and allege that an appeal to casuistry can resolve moral controversies in a pluralist, postmodern secular world. However, they never develop an account of how this secular casuistry would function. Traditional casuistry was built upon the analysis of particular cases and their resolutions within a concrete, content-full moral tradition. Within that tradition certain cases and their resolution could be regarded as paradigmatic for moral dilemmas. The difficulty with a secular casuistry is that there is no way to decide which cases are to function as the paradigm cases. Again, like the appeal to middle level principles, casuistry, shorn of a particular moral viewpoint offers no way to select the central paradigm cases to make the machinery run. Even if a secular casuistry could develop a set of paradigm cases, outside of any particular theoretical and cultural framework or content, there is still no non-arbitrary way to describe the moral dilemmas before us and choose which case should be the model to resolve it.

Often the proponents of casuistry or middle level principles argue that there is sufficient moral 'agreement' in contemporary culture to achieve resolution of moral controversies. One must be cautious however to discern the different possibilities and levels of 'agreement'. When men and women 'agree' on the resolution of a moral controversy that agreement can mean very different things. The agreement can simply be a superficial agreement on what course of action should be followed, or it can be a 'deep' agreement which involves not only the course of action to be followed but the reasons as to why the resolution should be chosen. One can have agreement on the choice of a course of action without agreement on the reasons for justifying the choice. Without agreement on the reasons for the choice of an action there is little hope to generalize beyond the case at hand.

One can think of people agreeing to discontinue artificial feeding and hydration to a PVS patient for many different reasons. Both Peter Singer and Cardinal Ratzinger may agree that such treatments should be withdrawn. The one reaches this conclusion because there is general disutility in continuing the treatment while the other thinks that continued treatment would be a burdensome, extraordinary means and the idolatry of life. Such disagreement about reasons will make it difficult to extend the agreement reached in the particular case. One cannot justify, in general secular terms, the imposition of content-full moral visions by appeal to moral terms such as 'solidarity', 'human dignity', or 'sanctity of the body'. Any argument must begin with some set of premises. The epistemological dilemma for contemporary ethics is that there is no way to know which set of premises with which ethics should begin.

The difficulty for secular bioethics is that physicians, patients, and biomedical scientists often meet as moral strangers: without a common starting point, a shared moral vision needed to resolve moral controversies. This is not to deny that some moral dilemmas are not resolved in situations where men and women share some moral framework. It is to say that in morally pluralistic context, one cannot assume others will share the same values and the same ranking of those values. There is no foundational starting point, in theory, middle level principles, or casuistry, which is not particular or arbitrary in some way. Each must have some moral content to develop resolutions to moral dilemmas. Yet the content is purchased at the price of universality.

The recognition of the postmodern predicament leads to a recasting of moral authority and the resolution of moral dilemmas. If one cannot appeal to a particular concept of God, or to a particular understanding of moral rationality, or to a particular understanding of nature, in order to ground bioethics, there is only one source left: the authority of moral agents. If one cannot discover an authoritative moral vision to ground moral judgments then one must appeal to persons as the source of moral authority. When we meet outside a particular understanding of morality, we have only each other to appeal to in order to resolve moral disputes and in order to frame the fabric of moral interactions. It is for this reason that one finds the salience in the postmodern world of such practices as free and informed consent, the free market, and limited democracy (i.e., governments that recognize robust rights to privacy, areas where content-full moral views of the majority cannot be imposed on those in the minority). Absent agreement on external moral standards the only moral justification possible for moral strangers is derived from the agreement of persons as moral agents.[2]

This is a recasting of the rational hope of the West that has sought to resolve moral controversies by appeal to reason. If one cannot discover a content-full moral vision which grounds moral authority, then one can only gain moral authority from moral agents. If one is interested in resolving dilemmas with moral strangers peaceably, without recourse to foundational force or coercion, with moral authority that can be justified in a general secular terms, then moral authority can only be derived from the agreement of persons as moral agents. This is not an appeal to the value of autonomy or freedom but a procedural appeal. One does not rely on reason to discover moral authority, but on the moral will to create moral authority. The necessary condition of mutual respect (the non-use of others without their consent) is not grounded in a value given to autonomy, liberty, or persons, but is integral to the project of controversy resolution when God, nature, and reason have not succeeded in transmitting a general content-full moral vision. To have a morality for moral strangers one needs only to refrain from using others without their consent and acknowledge them as agents who can agree or refuse to collaborate.

Appeal to mutual respect allows us to understand in general secular terms (those that do not depend on a moral vision) when force and coercion is justified. Moral strangers can use each other only when they act with commonly conveyed moral authority. Those who use others without consent loose a commonly justified basis for protest when they are met with punitive or defensive force. Limited democracies draw upon the morality of mutual respect to provide protection from and punishment for the unconsented-to use of persons (e.g., murder, rape, burglary) as well as to insure the enforcement of the contracts. In addition they are able to create, within the canons of consent, common endeavors such as the creation of a basic health care system (limited solidarity). Rights to privacy must play a crucial role in any secular state or large endeavor. It should be noted that rights to privacy are not celebrated because of any positive value assigned to such rights. Rather, they mark out the limits of plausible moral authority of the state to intervene in the peaceable consensual actions of individuals.

Again, bioethics provides telling examples of a cultural change. Bioethics has any number of procedural mechanisms, (e.g., advance directives, informed consent, institutional review boards, government commissions, health care ethics committees), that are designed to help men and women of diverse moral viewpoints collaborate. These are procedural solutions that enable health care institutions and practitioners to navigate through the plurality of moral commitments. However the secular tier will be devoid of content.

III. THE BODY

Modern high technology medicine has transformed expectations of health care. Much of the technological development in biomedicine touches directly on issues involving ownership of the human body. Technological possibilities have opened up discussions about the ownership of human cells and tissues [21] as well as the use of genetic material [9]. Each of these discussions has renewed fundamental arguments about the propriety of such interventions. As both the object of interventions and the source of material for such interventions the human body is the focus of many issues in biomedicine. How one understands the body will direct the way medical interventions are undertaken and what one thinks constitutes proper moral behavior with one's body.

Discussion of these issues, on the secular level, often becomes confused and intractable. People have different intuitions about proper moral conduct. Yet, while people may have strong moral convictions about proper and improper use of the body and bodily parts and materials, one will not be able to make out these claims in general secular terms. Unless people share the same ranking of moral values, convictions, or intuitions, moral arguments will not be compelling since the initial premises are not accepted.

How one acts towards one's body will be guided by one's understanding of moral life. The choices of others, about the uses of their bodies, will be governed by their own moral views. In a secular society, with competing views of the good life, there can be no canonical notion of how one ought to treat the body. While one may disapprove of certain uses of the body, as many hold the practice of prostitution to be immoral, there is no secular moral authority to condemn such actions. As long as proper permission is obtained and agreements are kept, there is no legitimate moral justification for secular governments to prohibit actions. In a secular world one may lament the behavior of others but lack the moral authority to prohibit such actions. Yet there continue to be legal prohibitions and regulations on the use of the human body. In many secular nations, for example, prostitution is outlawed as is the sale of human organs or some forms of reproductive medicine. Despite the limits of secular moral reasoning many ethicists and bioethicists act as if there were no epistemological difficulties in justifying their positions. Particular intuitions and assumptions are put forth as though they ought to be endorsed by all rational men and women. One need only compare the visions of bioethics put forth by its practitioners to know that rational men and women are often asked to endorse conflicting points of view.

An important philosophical question is why such prohibitions should exist in a secular culture. One finds an interesting clue in the general justification for the prohibition in *Organ Transplantation* [7]. In speaking to the rejection of commercialization the task force said that society's moral values 'militate against regarding the body as a commodity' ([7], p. 96). It went on to cite a report from The Hastings Center which said:

> The view that the body is intimately tied to our conceptions of personal identity, dignity, and self-worth is reflected in the unique status accorded to the body within our legal tradition as something which cannot and should not be bought or sold. Religious and secular attitudes ... make it plain just how widespread is the ethical stance maintaining that the body ought to have special moral standing. The powerful desire to accord respect to the dignity, sanctity, and identity of the body, as well as the moral attitudes concerning the desirability of policies and practices which encourage altruism and sharing among the members of society produced an emphatic rejection of the attempt to commercialize organ [donation and] recovery and make a commodity of the body ([13], pp. 3-4).

The view expressed here is that the body is intimately bound to our conceptions of ourselves seems true enough. The difficulty which enters in is when one tries to give content to notions such as 'personal identity', 'dignity' and 'self-identity', 'sanctity', 'solidarity'. Particular attitudes and value commitments shape the suppressed premises which lead to such prohibitions. The Report also is committed to a particular ranking of values as is reflected in the place given to 'altruism' in the Report and its recommended prohibition [24,25]. But the values are often simply asserted without argument. Moreover, it is simply assumed that these values are shared by all. There is no explanation of why altruism should be ranked the way it is, for example.

One way to understand the claims of The Hastings Center Report and the Report of the DHHS Task Force is to see them as what Alasdair MacIntyre would call 'bits and shards' of a broken tradition [18]. Appeals to 'sanctity', 'integrity', and 'altruism' are often the remnants of a once powerful, widely accepted Judeo-Christian moral vision in the West. This vision gave content and coherence to the type of claims made in these different reports. The sanctity of the body was grounded in its relationship to a Creator God. There were appeals to altruistic love at the center of the tradition. But it is not clear how, in a secular culture, one is to ground such claims, since fewer and fewer people have any belief in God, or understand the meaning of such terms. Indeed, in a secular society there is no way to distinguish, morally, the different uses of one's body. From giving one's body to another in marriage to the life of prostitution there is no way, on a secular level, to make out moral claims about each use of the body. One is left then with a secular level which is devoid of particular content. If one cannot constrain the behavior of others by moral argument then one cannot justify such constraints in law and public policy. One can only appeal to the moral authority conveyed by the consent of each person over his body and constrain the behavior of others so as not to violate this authority.

One's body must be respected as one's person. The morality of permission secures one's possession of one's self and one's claims against others who would use one's body or talents without one's permission. At the same time, with appropriate permission one can understand all types of arrangements and transactions between moral agents. Moral agents can freely enter into arrangements of special services or obedience. Various forms of indentured servitude – entering the military, the monastery, a religious order, or becoming a prostitute – are all relationships which transfer to others, in whole or in part, rights that one has over one's self. Outside of concerns about how agreements are made or kept, secular moral discourse must pass over such relationships in silence.

The secular state can only justify a morally neutrality and protection of its citizens from unconsented-to encroachments as well as the enforcement of recorded agreements. A careful examination of the secular tier of moral discourse makes it very clear that there will be no canonical vision by which to understand ownership of the human body. Absent a canonical moral narrative one must rely on the authority of moral agents. In the tier of secular moral discourse all moral language will be eviscerated of all content. This pluralistic tier is set in contrast to the second tier of moral discourse: the discourse of 'moral friends' which will have content and meaning which is impossible on the secular level. Through an understanding of the stark contrast between these two tiers of moral discourse one will be able to see the limits of the moral authority of large scale secular states.

The secular state provides a framework in which moral strangers, those with diverse moral points of view, can meet and cooperate peacefully. The moral authority of the state is limited but crucial to a secular culture. The limits of the moral authority of the state do not eliminate the state. The state becomes central to protecting the rights and exchanges of moral agents. It must punish those who unjustly take from others by force or deception. It must enforce agreements that have been freely made.

IV. MORAL FRIENDS

This analysis of the secular, morally pluralistic arena is not a destruction of the moral community. Rather, it is the recognition of the limits of moral reason and the importance of moral community; that is, exchanges which take place between those who share similar understandings of the moral life, values, and the sources of moral authority. Moral communities become the focus of substantive moral discourse. Such moral communities come in great variety. They range from those based on ideology (Marxism, Cambridge Liberals) to those based in God's revelation. Within communities one finds content-full moral discourse about the human body.

One such community which has reflected extensively on the proper uses of the human body is the Roman Catholic community. This community has a common moral sense (e.g., 'the sanctity of life') with moral principles (e.g., 'No direct taking of innocent human life') and structures of moral authority (e.g., confessors). Roman Catholicism has a particular view of the human person and the sanctity of the human body that provides a contrast to general, secular moral discourse.

Within the vision of this community the body is understood as "God's masterpiece in the order of the visible creation" ([20], p. 71-72). It has a glory not only here and now but it will "enjoy immortality in the glory of heaven" ([20], p. 71-72). Man is "not the owner of his body not its absolute lord". Man's power over his body is, although limited, direct ([20], p. 55-56). Indeed within this vision there is a series of principles and norms that direct how the body should be treated and used. These principles are derived from the proper intrinsic function of the parts of the body ([20], p. 53-54). In this view the moral limits of the use of one's body are limited by the 'natural finality' (purpose) of the member or organ. Man does not, for example, possess the power for acts of destruction or mutilation. However, in virtue of the principle of totality, man can give individual parts for destruction or mutilation when there is a good for the being as a whole (e.g., surgery). This view of the proper finality of the organs or members of the body is the foundational assumption in the Church's teachings on issues such as contraception and reproductive medicine.

Society's power over the body – indeed generally – is limited to the purpose and action of a community. It is the Creator, not the community, who gives man dominion over his body. Society cannot directly deprive a man of the right to use his body as he sees fit. However, man will have to answer to the Creator for the choices he makes ([20], p. 55).

In this tradition one also finds approval for compensation for body parts in an address given by Pope Pius XII to eye specialists in 1956. He argued that acts of donation cannot be viewed as a duty or obligatory acts of charity. Such acts are supererogatory not obligatory. In the same address the Pope approves of compensation for such actions. He said:

> Moreover, must one, as is often done, refuse on principles all compensation? This question remains unanswered. It cannot be doubted that grave abuses could occur if payment is demanded. But it would be going too far to declare immoral every acceptance or every demand of payment. The case is similar in blood transfusions. It is commendable for the donor to refuse recompense; it is not necessarily a fault to accept it ([20], p. 381-382).

So while the Roman Church's tradition encouraged altruism and acts of charity it does not see such acts as obligatory, or the banning of compensation as mandatory.

Within the moral vision and tradition of this community one discovers a well articulated understanding of the human body, its origin, its purpose, and the basis of proper authority over it. Within this framework one finds a coherence in the meaning and use of moral language which is absent from the secular tier of moral discourse.

V. CONCLUSION

With the continued development of biomedicine, attempts to regulate the use of the human body by the power of the state will continue. One thinks, for example, of the closing section of *Donum Vitae* [5] which argued that the state has the duty to regulate and control reproductive medicine according to the *particular* moral point of view articulated in the document. Many in a liberal democracy would not only find the views of *Donum Vitae* objectionable but they would also object to the use of state authority to enforce such content-full moral views. However, many of the same people want the state to prohibit the commercialization of body parts, for example, because of their *particular* moral views about the body and commerce. They do not seem to realize that they have no better justifications, in general secular terms, for their position than does the Congregation for the Doctrine of the Faith.

The particular views of the proper use of the human body are just that: *particular*. For a secular state there is no *moral* justification for the deployment of state authority to impose a particular moral point of view. Deployment of state authority has moral justification in procedural matters such as protection from coercion or the enforcement of agreements. To avoid the particularity of moral visions and ideologies the secular state will be libertarian by default. This is the only position which can be morally justified in general secular terms.

Kennedy Institute of Ethics
Department of Philosophy
Georgetown University
Washington, D.C., U.S.A.

NOTES

1. I am indebted to H.T. Engelhardt, Jr., mentor, partner, and friend, for the development of many of my views expressed in this essay. While I disagree with him in his understanding of moral community, I share many of his views on the limits of moral reason and the limits of state authority to enforce morality.
2. Certainly the author of this paper believes that objective standards of morality do exist. The conceptual difficulty is epistemological in that there is no 'view from nowhere' which enables us to know which are the correct standards. While philosophers have attempted to ground objectivity in the idea of 'supervenience' this seems, to the author, to be a secularized version of God's grace.

BIBLIOGRAPHY

1. Aquinas, Th.: 1948, *Summa Theologica*. Christian Classics, Westminster, MD.
2. Beauchamp, T. and Childress, J.: 1989, *Principles of Biomedical Ethics*, (third edition). Oxford University Press, NY.
3. Brody, B.: 1988, *Life and Death Decision Making*. Oxford University Press, NY.
4. Carse, A.: 1991, 'The 'voice of care': Implications for Bioethical Education', *Journal of Medicine and Philosophy* **16**, 5-28.
5. Congregation for the Doctrine of the Faith: 1987, 'Instruction on respect for human life in its origin and on the dignity of procreation: Replies to certain questions of the day', *Origins* **16**, 697-711.
6. Daniels, N.: 1985, *Just Health Care*. Cambridge University Press, NY.

7. Department of Health and Human Services: 1986, *Organ Transplantation: Issues and Recommendations*. U.S. Government Printing Office, Washington, D.C.
8. Engelhardt, H.T.: 1991, *Bioethics and Secular Humanism: The Search for a Common Morality*. SCM Press, London, UK.
9. Engelhardt, H.T. and Wildes, K. W.: 1991, 'The artificial donation of human gametes', in W.A.W. Walters (ed.), *Human Reproduction: Current and Future Ethical Issues* (Bailliere's Clinical Obstetrics and Gynaecology). Bailliere Tindall, London, UK.
10. Frankena, W.: 1973, *Ethics*, (second edition). Prentice Hall, Inc., Englewood Cliffs, NJ.
11. Gilligan, C.: 1982, *In a Different Voice: Psychological Theory and Women's Development*. Harvard University Press, Cambridge, MA.
12. Grisez, G. and Boyle, J.: 1979, *Life and Death with Liberty and Justice*. University of Notre Dame Press, Notre Dame, IN.
13. Hastings Center: 1985, *Ethical, Legal and Policy Issues Pertaining to Solid Organ Procurement: A Report on the Project on Organ Transplantation*. The Hastings Center, Hastings-on-the-Hudson, NY.
14. Hegel, G.F.W.: 1967, *The Philosophy of Right* (translation by T.M. Knox). Oxford University Press, NY.
15. Jonsen, A. and Toulmin, S.: 1988, *The Abuse of Casuistry*. University of California Press, Berkeley, CA.
16. Locke, J.: 1980, *Second Treatise of Government*. Hackett Publishing Company, Indianapolis, IN.
17. Lyotard, Jean-Francois: 1984, *The Postmodern Condition* (translation by G. Bennington and B. Massumi). Manchester University Press, Manchester, UK.
18. MacIntyre, A.: 1981. *After Virtue*, University of Notre Dame Press, Notre Dame, IN.
19. MacIntyre, A.: 1988, *Whose Justice? Which Rationality?*. University of Notre Dame Press, Notre Dame, IN.
20. The Monks of Solesmes: 1960, *The Human Body*. The Daughters of St. Paul, Boston, MA.
21. Office of Technology Assessment: 1987, *New Developments in Biotechnology: Ownership of Human Tissues and Cells*. U.S. Government Printing Office, Washington, D.C.
22. Pellegrino, E. and Thomasma, D.: 1981, *A Philosophical Basis of Medical Practice*. Oxford University Press, NY.
23. Singer, P.: 1979, *Practical Ethics*. Cambridge University Press, NY.
24. Spital, A.: 1991, 'The shortage of organs for transplantation: Where do we go from here?', *New England Journal of Medicine* **325**, 1243-1246.
25. The United States Congress: 1984, *National Organ Transplantation Act* [PL 98-507].
26. Veatch, R.: 1981, *A Theory of Medical Ethics*. Basic Books, NY.

PAUL SCHOTSMANS

OWNERSHIP OF THE BODY: A PERSONALIST PERSPECTIVE

I. INTRODUCTION

All who are familiar with recent developments in Roman Catholic medical ethics and moral theology in general recognize the importance of the methodological shifts which have taken place in the discipline. However one characterizes these shifts and the debates they have engendered, it is apparent that they have brought with them a number of issues which, while they played a secondary role in 'traditional' (physicalist, a priori, double-effect) medical ethics, are now emerging as central issues in their own right within a 'revisionist' (personalist, a posteriori, experiential, proportionalist) methodology. Personalism is based on the needs of the person, not on the structures of (biological or physical) nature. A personalist foundation for ethics includes explicitly both individual and social values [11]. One of the most important representatives of personalism, Louis Janssens, emphasizes that the human person "is essentially a social being: we must consider societal life as co-existence, co-operation, and co-participation" ([9], p. 223). Individual and social are interdependent in Janssens' personalism, but this has not always been clear in Catholic social and medical ethics. The history of the debate over organ transplantation within Roman Catholic medical ethics offers a profitable avenue for this kind of exploration: the principle of totality was, by the mid-twentieth century, an integral part of a physicalist and largely individualist moral methodology whose central framework was the doctrine of double effect. In this methodological context, the notion of totality had come to be limited to that of the physic good of the individual physical organism. After the radical reforms introduced by Bert Cunningham ([11], p. 153; [4]), nuanced by Richard McCormick using the notion in the direction of the total personal good of the individual donor, organ transplantations were justified by appealing to the principle of totality since the donor may rightly subordinate his or her own physical perfection to his or her own spiritual and personal good [15]. As Louis

H.A.M.J. ten Have and J.V.M. Welie (eds.), Ownership of the Human Body, 159–172.

Janssens would repeat the position of Vatican II, the human person is an embodied subject and as such is essentially relational, that is, open to the world, to other people, to social groups, and to God (*Gaudium et Spes*, nr. 51). The insistence by most moralists of the period on the physicalist and individualist limitations of the principle of totality and, more importantly, their insistence on physicalism and individualism generally in medical ethics, is an approach that most of today's Catholic moralists have rejected. With David Kelly and Louis Janssens, I agree with this rejection. My own preference is an ethics of responsibility built upon a personalist foundation in the spirit of the Second Vatican Council (Pastoral Constitution on the Church in the Modern World: Gaudium et Spes). This essay will deal with the anthropological foundations of this position and with the ethical clarification of the human body from this perspective.

II. AN ETHICAL-ANTHROPOLOGICAL CLARIFICATION

Modern developments in medical technology have the virtually unique character of confronting individual human beings with their personal responsibility. The reality of the human condition is much broader than what some may consider the absolute reign of freedom: humans are situated, incarnated beings. By their very being they are involved, fully relational and inter-human. To absolutize their freedom would mean to dispose of the most human possibilities: openness unto oneself, to others, and the Other. Freedom and, in its line, the ownership of the human body has to be integrated in an anthropological description of the humanly desirable. I will therefore first illustrate that the reality of responsible involvement is basic to an adequate conception of freedom.[1]

From Unicity through Inter-Subjectivity toward Responsibility in Solidarity

In the search for the core of our being human, we are best served by recalling part of the history of anthropological reflection. What we find striking about that history is that one virtually can draw a straight line from the source of this systematic reflection on humanity, which more or less coincides with the beginning of the twentieth century, up to the present day, from the recognition of the openness of being human to the importance of social responsibility. This observation may help us to avoid falling into one-sidedness. What we shall try to do here is to suggest a complementary vision on humanity and on being human. For it appears to us to be a dead end simply to stress one or another aspect of what it means to be human.

According to this vision, to be human is both to exhibit and to participate in the wonder of a rich many-sidedness: to be human is to be rich in unicity and originality, but at the same time, originality is an empty concept if it does not include openness toward the other and if it does not involve cooperation with others for the exploration of a solidary community.

a. Unique and Original: the Person at the Center

There was a period in which little if any attention was given to the mystery of the unique human being. Attention was focused mainly on human knowledge or, more broadly, human consciousness [18]. However, the radical experience of the First World War opened the eyes of many European philosophers. Suddenly, one discovered the unique and concrete human person with his or her own life-project, his or her fears and expectations.

The being-human of each unique person is an attempt to realize oneself in freedom. Typically, the individual constructs the self with a view toward meaningful existence. With the knowledge of what one is and what one can become, one may well put everything into the service of one's own life project [14]. Here lies the contemporary anthropological foundation of human labor: the individual makes the self a project and puts everything in motion to realize this project. Furthermore, the initial insight clearly saw that every human being is different, a unique and proper I, a person with talents, capacities, feelings and possibilities. In interaction with social-cultural surroundings, everything came to be integrated toward the formation of a unique, original personality with an individual character. Ultimately, this led to the idea that the human being stands in relation to him- or herself, and that it is from this point that one begins the developmental process.

This discovery of the unicity of being human was so profound that in the beginning it was formulated absolutely and radically. This might explain how things have gone awry, moving toward an egotistical fixation on the concept of the human. Nevertheless, the reflection on what it means to be human did not stop there. Existentialist philosophy – as this movement is usually called – has had a great influence on the spirit of our age. Yet the one-sidedness with which the unicity of each human being was stressed itself virtually led to the appearance of a counter-movement. In medical ethics, it has only very recently been realized that it would be pernicious to base ethical reflection solely on the right to self-determination, as this concept of the human person is frequently translated.

On the one hand, we should not fear voicing an appreciation for the fundamental insight of existentialism: the unicity and originality of each personal experience of being human. On the other hand, however, it appears necessary to search for a complementary approach, one which is not limited to the unique, but isolated, experience of being an individual.

b. Relational and Inter-Subjective: the Human in Dialogue with Others and with God

The exclusive affirmation of the unicity of the human person could prevent us from seeing the wonder of being human in its multi-sidedness. To grasp the whole of this mystery, we must consider as well the openness of each human being toward fellow-humans. As early as in 1923, the Jewish philosopher, Martin Buber, wrote his pioneering work on being human, *I and Thou* [3]. With this most valuable contribution, a new insight broke through: one can never be a human being alone. As humans we essentially stand in an open relation, involved with the reality in which we live, with other humans to whom we owe our existence and who continue to surround us, and ultimately with God. If humans wish to be fully themselves, they stand in need of encounter with others and of being encountered by others. It is also important to consider the distinction that Buber made between what he called the *social* and the *interhuman*. To the *social* belongs everything that can be described as the being together of a multitude of humans, i.e., the experiences and reaction patterns that are connected with coexistence and cooperation. Over against the oppressive collectivity of these cooperative relations, Buber presents the interhuman as the sphere of the immediate, of the life of person-to-person. The dialogical is therefore the full unfolding of this atmosphere.

Buber attempted to capture the dividing line between the social and the interhuman by making a distinction between the 'I-it' and the 'I-thou' relations.[2] I-it mainly concerns the businesslike, calculating relation with the reality in which one is situated. In this world, everything turns upon having and dominating, owning and using. The I-Thou is the personal, true relation of the human with reality. The I confronts a Thou in the fullness of the person. There is a complete *immediacy* in the relation with the Thou, which can be a human being, a spiritual entity, or an element of nature. Only here does one grow into whom one really can be. Therefore, this relation can be fully characterized as *encounter*: the singularity of being human is illuminated by the immediacy of this encounter between I and Thou.

To that end, Buber refers back to a notion that is even better suited to break through every human self-closed-ness: the *between* as the primal category of the human mystery. In this between – in which the Thou of the other is immediately revealed and in which the I can immediately participate in the being of the other – the becoming-a-person of every human reaches its climax. Thus, the human person only becomes human thanks to the Thou and because of the Thou. No one can become human through self-concern, limited only to oneself. One becomes human through the other-than-oneself.

With the concept of the *between*, Buber is opening the gates widely for a theological reading: "prolongated, the lines of relations converge in the eternal Thou; each special thou is a perspective on the eternal Thou"' ([3], p. 91). What is most striking here is that Buber elevates the human being to a full fledged partner with God. In this sense, he opposes the absolutization of every feeling of dependency. It should be added that the eternal Thou can never become an It, because it is essentially a Thou.

The limitation of this dialogical philosophy mainly lies in the fact that it overstress the small-scale and the intimate encounter. By this, it is as it were reserved for the silence of the profound discussion, the mystery of the revealing togetherness between two persons. However important this may be, this dialogical philosophy is only valuable if it is supplemented by an analysis of the unicity of the being of every human (cf. supra) and of one's solidary responsibility for a just society (cf. infra).

c. Communication and Solidarity

The insight into the fullness of being human is again enlarged by the notion of participation in the whole of coexistence. We refer here to the phenomenon of living in a particular, concrete society as such and the ethical assignment that accompanies social living for realizing the *good life* [6]. The investigation of the mystery of our being-human here encounters a new, fascinating discovery. For a long time, social commitment remained outside the scope of the majority of anthropological reflection. It is not surprising that this insight broke through because of the contrast experience of human beings confronting a social order that was largely inhuman. The excesses of the industrial revolution (in the preceding century) and of the dog-eat-dog relations between totalitarian states (in this century) functioned like a scream, heard by those who were committed to humanity. Therefore, it was primarily the victims of social desperation who gave voice to the task of solidary responsibility.[3]

This might explain why those who accept such an approach go about designing an ideal image of a just society. Yet, no matter how one presents it, it will never be possible to fabricate a perfect society. The most one can hope for is a movement toward the approximation of an ideal of justice, to come as close to it as possible. It will always remain possible to point out imperfections. Or, to express it in the terms of Emmanuel Levinas, even in the best welfare state the public servant does not see the tears of the individual ([12], p. 102-103). And so we come to the implied concept of permanent revolution which is so particular to this approach: again and again what has already been achieved, or the already existent situation, continuously needs to be questioned and reoriented toward the more humane. The biblical condemnation of building images here takes on a new significance: the truth about humankind and society can only be understood as a permanent process of critical transcendence of the already existent and the already accomplished: *societas semper reformanda*.

Theologically, it has been mainly political theology that has made this shift in anthropological questioning. We would draw attention to a model that has proven to encompass a greater ethical range and is therefore more usable, namely the model of communicative ethics of Karl-Otto Apel [1]. From what we said above, it has already become clear that for the first time in the history of anthropological reflection, human beings have been explicitly placed before the task of assuming solidary responsibility. Individual and relational ethics are no longer sufficient. But still, the complementary construction of our approach should guard us against premature conclusions. Thus the central question remains: how can the personal decisions of conscience of a unique person express and integrate this connection with a solidary responsibility for society ([13], p. 40)?

With this model of the communicative ethics, Apel is more or less trying to reconcile these two poles. He therefore reflects on the way that people communicate with each other. In speaking, taking a position, or defending a line of argumentation, one always does this in respect to others. If not, one's argumentation makes no sense. Even one who attempts to go about reflection entirely on one's own cannot escape from this. One who reflects can only articulate and verify thoughts in the form of a dialogue, even if it is a dialogue of the soul with itself ([6], p. 95). All creatures that can speak therefore have to be introduced into the conversation – the ultimate justification of thinking can exclude neither a partner nor any potential contributions from participants in the discussion.

The coexistence of humankind is thus always an existence in communication. The ethical norm behind all this demands that not only assertions but also the claims of people over and against people must always be justified in dialogue. Here one arrives at the foundational demand that "not a single, limited individual interest of a human being may be sacrificed" and the ethical principle that "all needs of all people – as virtual claims – must become the concern of society, at least insofar as they can be brought into accord with the needs of all the others" ([1], p. 425).

Therefore, someone who is formulating an argument, thinking, speaking, always postulates two things at the same time: the fact that there exists a society around him or her of which he or she is a part, and the hope and the expectation that everyone else can understand what is brought to the fore. At the same time, this offers us two fundamental principles for life in society: first, social life must concern us, in everything we do or omit, to secure the survival of humankind in society, and then to continue travelling the road toward a more just society. It is clear that the image of 'blessed future,' of which we can only dream, is finally determinative of the way in which we are willing to let people today exercise their rights. To say it in another way: the second principle (the image of a perfect society) determines the content of the first (concrete solidarity with the whole of humankind).

The Personalist Criterion

It seems to be important to demonstrate how responsible involvement as an important aspect of the interwoveness of personal self-development, relational encounter, and human solidarity can deal with new possibilities of medicine. This approach provides an ethical as well as a pedagogical frame of reference. From this perspective, the responsibility of every human being for the meaningful integration of these dimensions in his or her life project can be taken up. This is the ultimate task of ethical reflection: to offer the opportunity to each person to contemplate his or her own situation in a sufficiently rational way, to become conscious of all the values and disvalues that are present to us as persons, and to open our perspective on the humanly desirable.

In accord with this criterion we say that an act is morally good if it serves the humanum, that is, if in truth – according to the three dimensions – is beneficial to the human person adequately considered in himself and in his relations. In virtue of the historicity of the human person this criterion requires that we again and again reconsider which possibilities we have at our disposal at this point in history to serve the promotion of the human person. That is a demand of a dynamic ethics which summons us to the imperative of what is better or more human according as its actualization becomes possible. In conjunction with this we must, in our acts, respect the originality of all as much as possible [10].

We will now examine more deeply the dimension of corporeality as a basic orientation of the human person and the humanly desirable.

III. A HUMAN SUBJECT IN CORPOREALITY

From a personalist perspective we consider the person as a subject in corporeality. That we are corporeal means in the first place that our body forms a part of the integrated subject that we are: corporeal and spiritual, nonetheless a singular being. What concerns the human body, therefore, also affects the person himself.

That we are a subjectivity, or a conscious interiority, in corporeality (Gabriel Marcel: *un esprit incarné*), is for Marcel, the basis for a number of moral demands. Usually, we take care of our own health and bodily integrity, as well as that of others. We take account of the limitations of our bodily strength, and do not usually consider our bodily needs and tendencies merely as biological givens. In this context, the Vatican council remarked that "the sexual characteristics of man and the human faculty of reproduction wonderfully exceed the dispositions of lower forms of life" (GS, 51).

Our body forms not only a part of the subject who we are, but also, as corporeal, a part of the material world. Through this very fact, our being is a being-in-the-world. Because we are corporeal, we need the things of the world. And in order to put these things in service to our needs, we must, by our work transform the world from a natural milieu into a cultural one. We must make the world continuously more liveable for human persons. "For man, created in God's image, received a mandate to subject to himself the earth and all that it contains", and by his "labor to further unfold the work of the Creator" (GS, 34). With the aim of improving the world "the sciences take on mounting importance" and "technology is now transforming the face of the earth and is already trying to master outer space" (GS, 5).

Nevertheless, we should not here lose sight of the ambiguity of our human creations. In the first phase of scientific and technological advancement we were especially impressed with its positive contributions: the growth of production and prosperity. At present, however, we are very much concerned with the negative effects which inevitably appear to be bound up with this. We are beginning to feel very deeply that the purity of air, water and land and the control of traffic and noise are questions about the life of subjects in corporeality and that the protection of the environment is a moral obligation. Even more subtly, we have become aware that this productive and consuming society is influencing us so onesidedly with respect to what is prolific and useful that we are threatened with becoming one-dimensional beings and forgetting that our bodily-being-in-the-world embraces thinking and feeling as well as seeing, hearing, tasting, smelling, and touching, and therefore also involves the enjoyment and appreciation of things as well as aesthetic wonderment and contemplative reflection upon the deepest meaning of things and persons.

Just as in the domination of nature in general, dealing with human infertility, the manipulation of procreativity and the treatment of the dying have an extremely strong technological strain. This is certainly true for all the techniques applied to, e.g., artificial insemination, fertilization in vitro, artificial respirators, nutrition and hydration, organ transplantations etc. This is true as well, or perhaps more so, for future interventions which scientists foresee not as fiction but as realities in the more or less immediate future. As these things develop, one will have to take account of the ambiguity of human activity at every turn with its positive and negative aspects.

What personalists intend to emphasize is that an ethics of responsibility on a personalist foundation does not remain caught in subjectivism, but rather substantially holds to objective criteria: the human person adequately considered in his essential aspects or constant dimensions. Critiques are, however, essentially directed to the ways how this general principle can be of use in the concrete praxis of our moral life, exactly the place where our corporeal existence is functioning at the most intense level.

IV. PERSONALISM IN DISPUTE

Personalist ethic clearly uses what can be called a proportionalist method in the development of ethical reasoning. As being described by Gaillardetz: "Catholic moral theology has over the last twenty years seen the development of a 'revisionist school' whose approach to central ethical issues is grounded in a methodology often know as 'proportionalism'. This methodology has gradually found its way into the mainstream of Roman Catholic moral theology where, in varied forms, it now has achieved wide acceptance" ([8], p. 125).

Personalist ethic disputes clearly the view that Christian morality, although based on the revealed Gospel, must also conform to 'the natural moral law', accessible to human reason by reflection on our universal experience of our common human nature. The legalism of the deontological system of ethics has been strongly criticised by the influential Bernard Häring in many of his works. Josef Fuchs, in an important article in 1971 [7], proposed that the notion of 'value' might be used to revise the traditional physicalist Principle of Double Effect.[4] According to that principle, when an act entails both good and bad consequences it is necessary to meet certain criteria before it can be judged ethical. The first of these criteria is that the act not be intrinsically immoral, and the second is that the good consequences be proportionate (greater or equal to the bad consequences). Fuchs, without rejecting the principle, suggested that this first condition was unnecessary since in every case what makes an act morally evil is not the act considered in the abstract but whether in the concrete circumstances in which it is performed there is a proportion between the values and the disvalues which an act embodies. He argued that such a method is more objective than the traditional method since it takes into account all the values and disvalues involved, and he drew the conclusion that although there are 'absolute' values expressed in general norms, in the concrete application of these norms there may always be (at least in theory) room for exceptions. This system has become widely influential and in practice means that its proponents defend the commonly accepted norms of Catholic morality, but contend that in difficult cases exception should be made to these norms when a careful weighing of the values and disvalues warrant it.

The critics of this theory have raised many objections against it, to which its defenders have attempted replies. In particular the defenders deny that they believe that it is ever permissible to perform an evil (immoral act) to accomplish a good end. Louis Janssens, e.g., adopts the distinction between an 'ontic' or pre-moral evil and a moral evil. For example, homicide is a negative ontic value, which can become either a negative moral value (murder) or a positive moral value (self-defense) in view of the circumstances and intention of the agent. In judging the morality of an action, the ontic values and disvalues to be weighed are not in themselves either moral or immoral. Morality enters only in the proportion of values to disvalues. Consequently, an act in which the values exceed the disvalues is by definition never an immoral or evil means even if it is contrary to some generally valid norm.

Benedict M. Ashley vehemently criticizes this position. He refers to critics of this proportionalism that argue that in practice it comes down to utilitarianism. Ashley himself rejects this moral method as contradictory: "Proportionalism seems to me to be only one of a number of current proposals for a revision of traditional Christian ethics ... which are based on the belief that modern science has rendered obsolete the notion of a human nature which can be empirically explored, and the cultural sciences have made it necessary for us to accept cultural relativism" ([2], p. 369).

Personalist ethicists stressed essentially the importance of the teleological methodology, referring to the promotion of the humanum, that is, what is beneficial to the human person adequately considered in himself (as personal subject in corporeality) and in his relations (in his openness to the world, to others, to social groups, and to God). The human person is the ultimate goal and the dynamic criterion. Accusations of utilitarianism and relativism are therefore inadequate and unjust. One of our main objectives is to eliminate from Catholic ethics what Kelly and Curran have called 'physicalism' or 'biologism' [5]. This implies at the same time that they reject a naturalistic theology of the body.

As all Catholic moralists, also the personalist ethicists repudiate utilitarianism because in its act version (cfr. Joseph Fletcher) it repudiates moral norms altogether (except perhaps some single norm such as 'act lovingly' or 'the greatest good of the greatest number'), while in its rule version (accepted by most utilitarians) it reduces all moral values to a quantitative calculus of good and bad consequences (consequentialism), a calculus that seems inapplicable to the spiritual values given highest rank by the Gospel.

What personalists emphasize is that an ethics of responsibility on a personalist foundation does not remain caught in subjectivism, but rather substantially holds to objective criteria: the human person adequately considered in his essential aspects or constant dimensions. The corporeal and material dimensions are basic for this understanding. All human actions are indeed characterized by an ambiguity. This is true for two reasons. The first is our temporality. As we, at a given moment, choose to perform a certain action, we must simultaneously neglect all other possibilities, at least temporarily. The second reason is our spatiality. Janssens understands hereby that our actions, as active commerce with material realities – our own body, the things of the world – with their complex multiplicity of properties and physical laws involve an inseparable connection between negative aspects (disvalues) and positive attributes (values), such that they are simultaneously both detrimental to and beneficial for the human person: when are there proportionate reasons to perform an activity in a morally responsible manner which simultaneously results in values and disvalues ([10], p. 16-17)?

V. CONCLUSION

The ownership of the body is fundamentally an anthropological issue. We propose an anthropological model that keeps open the basic interwoveness of our being human. The medical problems as such are not our main concern. It seems to us to be much more important to illustrate how responsible involvement as an important aspect of the interwoveness of personal self-development, relational encounter, and human solidarity can deal with new possibilities in medicine.

This is the ultimate task of ethical reflection: to offer the opportunity to each person to contemplate his or her own situation in a sufficiently rational way, to become conscious of all the values and disvalues that are present to them as human persons, and to widen the perspective on the humanly desirable. At the same time, this approach contains an alternative to possible abuses and one-sided options. Finally, this anthropological model keeps open the possibility of exercising our essential solidarity.

School of Medicine
Catholic University of Louvain
Louvain, Belgium

NOTES

[1] See also [16].
[2] See [3], 9th edition from 1977, p. 9.
[3] Here special mention should be made of Theodor W. Adorno and Max Horkheimer, both members of the Frankfurt School.
[4] He did so under the influence of the writings of the German philosopher, Max Scheler [17].

BIBLIOGRAPHY

1. Apel, K.O.: 1973, 'Das Apriori der Kommunikationsgemeinschaft und die Grundlagen der Ethik', in *Transformationen der Philosophie*, II. Suhrkamp, Frankfurt a.M., pp.358-436.
2. Ashley, B.M.: 1985, *Theologies of the Body: Humanist and Christian*. Pope John XXIII Center, Braintree, MA.
3. Buber, M.: 1923, *Ich und Du*, (9th. ed.). Schneider, Heidelberg.
4. Cunningham, B.: 1944, *The Morality of Organic Transplantations*, Diss., Catholic University of America Studies in Sacred Theology, 86, Washington Catholic University of America Press, Washington, DC.
5. Curran, C.E.: 1977, 'Utilitarianism and Contemporary Moral Theology: Situating the Debates', *Louvain Studies* **6**, 239-255.
6. De Clercq, B.J.: 1980, *Menselijk samenleven als opdracht. Grondlijnen van een sociale ethiek*. Acco, Leuven, Belgium.
7. Fuchs, J.: 1971, 'The Absoluteness of Moral Terms', *Gregorianum* **52**, 415-458.
8. Gaillardetz, R.: 1989, 'John Finnis and the Proportionalism Debate: A Critique of a Critique', *Louvain Studies* **14**, 125-142.
9. Janssens, L.: 1967-77, 'Norms and Priorities in a Love Ethic', *Louvain Studies* **6**, 207-238.
10. Janssens, L.: 1980-1981, 'Artificial Inseminations: Ethical Considerations', *Louvain Studies* **8**, 3-29.
11. Kelly, D.F.: 1988, 'Individualism and Corporatism in a Personalist Ethic: An Analysis of Organ Transplants', in J.A. Selling (Ed.), *Personalist Morals*. University Press, Leuven, Belgium, pp.147-165.
12. Levinas, E.: 1962, 'Transcendance et hauteur', *Bulletin de la Société française de philosophie* **56**, 89-113.

13. Luijk, H. van: 1979, 'Karl-Otto Apel en de grondslagen van de ethiek', *Tijdschrift voor Filosofie* **41**, 35-67.
14. Macquarrie, J.: 1980, *Existentialism*. Penguin, Harmondsworth, UK.
15. McCormick, R.A.: 1975, 'Transplantation of Organs: A Comment on Paul Ramsey', *Theological Studies* **36**, 503-509.
16. Schotsmans, P.: 1988, *En de mens schiep de mens. Medische (r)evolutie en ethiek*. De Nederlandsche Boekhandel/Uitg. Pelckmans, Kapellen, Belgium.
17. Scheler, M.: 1916, *Der Formalismus in der Ethik und die materiale Wertethik*. Niemeyer, Halle, Germany.
18. Strasser, S.: 1963, 'Fenomenologieën en psychologieën', *Algemeen Nederlands Tijdschrift voor Wijsbegeerte en Psychologie* **56**, 1-21.

UFFE J. JENSEN

PROPERTY, RIGHTS, AND THE BODY: THE DANISH CONTEXT

A Democratic Ethics or Recourse to Abstract Right?

I. INTRODUCTION

Some years ago a clergyman tried during a discussion on medical ethics to focus on some main problems in the Danish health care system by offering the following comment: 'If the doctor takes even the smallest of the patient's personal belongings when he or she is dead, the doctor will immediately lose his job. If the doctor, however, takes the patient's body or parts of it for his research nothing at all happens. On the contrary, this is accepted as the most natural thing in the world, and this in spite of the fact that our body belongs to us in a much more essential way than (let us say) a watch'.

In the meantime, considerable changes have taken place in the Danish health care system, partly as a result of the ongoing and still more widespread debate on problems of medical ethics. We have new laws regulating, e.g., organ donation, autopsy, etc. We *apparently*, to a much greater extent now than before, take into account that the body belongs to the individual patient or citizen, as his or her personal property.

I will present some examples to illustrate this development and address the following question: "Are the changes really based upon a new recognition of the body as the individual being's personal property?" I will discuss to what extent it is reasonable or justified to consider the body and its parts as personal property, and, furthermore, what might be the shortcomings or unintended consequences of such a conception or theory.

The clergyman's critique of what was some years ago the practice concerning autopsy should be interpreted as a critique of an entire practice or policy. We can distinguish between different policies for obtaining cadaver organs; policies of 'giving organs', 'taking organs', 'trading organs', and 'selling organs'.

The Danish policy was a policy of taking the body away for autopsy and of taking organs from the body before it was 'delivered' back to the family since this was considered justified since it advanced medical

H.A.M.J. ten Have and J.V.M. Welie (eds.), Ownership of the Human Body, 173–185.

research. Autopsy and removal of organs would usually be performed under conditions that would not burden the bereaved family. It was well-known that autopsy (and removal of organs for research purposes) was performed, but only a few donors or their families objected to the practice (which took place within 8 hours after the patient's death had been certified).

Now, according to legislation of 1988, *consent* from the patient or the family is a necessary condition for performing autopsy. A number of medical specialists have warned against the law, claiming that a decline in the number of bodies autopsied would have serious consequences for medical research and medical education.

II. BRAINS IN A VAT

After passing the new Bill, a particular case has provoked public controversy between supporters of the old practice and supporters of the new regulations. It has come to light that thousands of brains of deceased psychiatric patients have been stored at the Psychiatric Hospital in Aarhus.

To some, this shows the cruelty of the former practice, for others it illuminates the medical and ethical rationale of this practice; through further brain research we will obtain further knowledge of the biological mechanisms behind and causes of psychiatric diseases. It is argued that the brains stored at Aarhus will be an invaluable resource in the ongoing and even more intensive research in biological psychiatry. Some of the critics demand that the brains be buried to show respect for the patients whose brains were removed *without their consent* (though in accordance with the old law). Others reject this as a futile gesture which would destroy a source for research which might prove to be of inestimable value to future psychiatric patients.

This controversy might at first be interpreted as a conflict between a utilitarian and a deontological ethics, the latter taking autonomy or integrity of patients as a basic and unconditional value. This would, however, amount to considering the case in isolation from its medical-cultural context and the values implicit in this context.

The traditional procedures for autopsy are imbedded in a disease-oriented practice, the basic intention of which is the saving or prolongation of life. As long as anything can be done for the benefit of individual patients, it is the doctor's primary duty to take care of his patient. The individual patient should be considered as an *end*, but when the patient has died the experience and knowledge acquired through treating the patient should be used as a *means* to improve the treatment of other patients and so, potentially, to contribute to prolonging *their* lives. However, from the perspective of disease-oriented practice, clinical experience is more than the experience acquired in the treatment and analysis of the patient when he or she is alive. Some aspects of the disease can only be disclosed through autopsy. It is part of the doctor's duty to reveal such medical facts about cases he has treated in order to treat future patients [3].

From within this practice, based on the clinical-cultural value of beneficence (related to the clinical-cultural value of prolonging human life), it is simply incomprehensible, ethically unjustified or even absurd, to demand that the brains stored at the Psychiatric Hospital in Aarhus should be buried. It is to deprive future patients of better treatment which should be (and, if available resources are used, *could* be) offered by the disease-oriented practice.

III. CONFLICTING VALUES

The new Danish Bill demanding consent as a condition for autopsy is only one of several examples which illustrates that cultural consensus concerning disease-oriented practice and its basic value is no longer paramount. It is, however, not at all clear what the main reasons are for the change and there is no clear-cut picture of the ethical perspectives which counteract the values embedded in disease-oriented practice.

The principle of autonomy has been cited by critics (philosophers and others) against the paternalism which has been, and still is, a dominant feature of the Danish health care system. It is, however, not at all clear if autonomy is a genuine concern of the public in matters of medical treatment and health care. Patients complain that in many cases they are not being taken seriously by doctors, that doctors don't listen to or inform them properly. But it is not clear if patients in general demand recognition as agents (and, if they do, in what sense they actually understand personal autonomy) concerning the planning and administration of medical treatment. During an innovative project, which was initiated at the cancer ward of Aarhus City Hospital – that involves patients in specifying ends (targets) for treatment and the ongoing evaluation of treatment and care – patient representatives were reluctant to include *decisions for treatment* in the project. Some patients explicitly expressed the fear that things would 'go wrong' if other than medical experts were involved in clinical decision-making procedures.

But if a new ethical discourse that focused on patients' autonomy had not been the catalyst in the process which is radically changing the health care system, then what can explain this change?

IV. A RHETORIC OF PROPERTY

Patient-frustration and negative experiences in the daily transaction with experts is hardly sufficient to explain the political willingness to implement these reforms, taking into consideration the strong resistance from professional groups and their organizations.

The rhetoric of property, i.e., a way of arguing which takes recourse to *property rights* rather than using traditional ethical arguments, seems to have played a role in the process. The clergyman's critique of the old autopsy-practice is just a symptom of the rhetoric of property or suspending ethical discourse. In an historical perspective it is not at all surprising that property-claims have come to play a role in an ideological climate characterized by critique of expert-power and medical paternalism.

The situation can be compared with the historical transition from feudalism to modern bourgeois society. Traditional values and ethical perspectives were overthrown in a socio-historical process where property-rights were espoused to legitimize the abandonment of the old order.

But why in the present situation talk about a *rhetoric of property* rather than a *new ethical theory or perspective* stressing the importance of recognizing the body as personal property?

The main reason is that we have no such theory. The political philosophers of the Enlightenment and their imported ethical theories assigns a cultural *role to* the concept of property and to the idea of the body as a person's property. This is quite another context than the present one where we face ethical problems germane to organ donation, genetic screening, etc. Or to put it in another way: if we already had an adequate ethical theory that focused on the person as owner of his body, we would not be as bewildered as we are, when we address the question of property-rights over organs, genes, etc.

Consider what makes the present discussion so complicated:

First: the question of how to approach the ethical question from the perspective of property-rights is confounded with the philosophical-anthropological debate about the concept of *person* (including the question of what constitutes a person and whether or not a person owns his body etc.).

Second: we have already in Hegel's philosophy a systematic and impressive argument to the effect that the basic ideas of property, property-rights, and person as a possessive individual cannot provide us with a suitable framework to account for the moral and ethical problems that face us.

V. PROPERTY AS RIGHTS

Property is not a special kind of thing. Properties are not things that belong to a particular ontological category. Properties are *rights in or to things*, and relations, society, and its particular institutions, through their social and legal procedures define, *some* physical possessions as properties. This means that to have property is to have a 'right to' in the sense of an enforceable claim to some use or benefit of something [6]. That properties consist of actual rights (in the sense of enforceable claims) in a particular society does not imply that they are immune to ethical assessment and critique. Social critics, philosophers, and others often have criticized rights or claims to properties in particular societies, by recourse to ethical theories.

When the rhetoric of property is removed from the ethical context it is natural that attempts will be made to ground the rhetoric in speculative metaphysics (or in elementary conceptual truths presented as if they were metaphysical truths).

Apparently, such a move might also seem to be a way to escape the *ethical relativism* of our age. Property claims have an ontological ring. Either a thing is my property or it is not my property, and is it reasonable to assume that those things which are properties are properties due to specific 'ontological' characteristics. But if property captures a set of rights, and if rights are rooted in social institutions and always have to be justified or assessed from the perspective of a given social institution or ethical perspective, then 'to be my property' does not seem to be a 'real' property in the objective world, in a metaphysical realist sense of real [7].

VI. AN ONTOLOGICAL GROUNDING OF PROPERTY CLAIMS?

Confronted with this challenge, the ontological or essentialist move may seem to be the only way to secure the objective status of property claims. The clergyman's claim that my body belongs to me in a much more essential way than does my watch seems to presuppose such an *ontological* grounding of property claims.

We might try to find a foothold for a theory of the ontological grounding of property claims using our intuitive distinction between things of which we are temporarily in possession and 'real' property. If I buy a concert ticket, I may say that 'my' seat is *in* my possession during the concert. Of course, I do not by that mean that the seat is my property. In a similar way, I can say about a book which I have borrowed from a friend but which is not my property, that it is (temporarily) in my possession.

We might face the objection that my claims to property simply mean that I have the right of use or benefit from something, which we call 'my property', in a way in which I do not have the right of use or benefit when it is a case of 'temporary possession'. I might accept this relativizing of the distinction between *properties* and *possessions* in the following way: It is true that in both cases I am related externally to an object (say 'my house' and 'my seat in the concert hall'), but, I might add that there are things which are my property in quite another and ontologically more basic way, things to which I am not related externally, but internally, things which essentially belongs to me. My body and my consciousness are entities of this kind.

In that light, the philosophical question of personal identity becomes relevant to the problem of rights and properties. If it, e.g., can be shown that I cannot be identified as the person I am, independently of my having a body in space and time, then the body seems to be mine in an *essential* way. I am as a particular person related intrinsically to that body. So it is not just something of which I am temporarily and accidentally in possession, but something which is my property in an ontological sense.

Now, let us assume that an analysis of the concept of 'person' (as the one just sketched), is correct, which it might well be, does this show or prove anything about possessions, or properties as rights in or to things?

Not at all. *A priori* this analysis would exclude ordinary things (e.g., my bike) from being real property, as such things are not conceptually, intrinsically, or essentially mine. By trying to ground the concept of property in *essential or necessary relations*, it is from the very beginning evident that this could not possibly be a way to explicate our practical concept of property, because this concept is a concept of relations (expressed in enforceable right-claims) which are subject to historical change and hence not necessary or internal.

VII. THE RHETORIC OF PROPERTY AS AN ARTICULATION OF GROUP INTERESTS

An ontological move cannot make the rhetoric of property a theoretically grounded claim. Instead of trying deceptively or self-deceptively to do that, a pragmatic rhetoric of property should be presented as what it is: a particular way of articulating the interests of individuals or groups who are questioning or fighting claims to rights in or of things put forward by authorities, powerful groups, or individuals.

In 1973, women in Denmark obtained a right to cost-free and unrestricted abortion (within 12 weeks after pregnancy). One of the arguments which has been used by feminist activists (in Denmark and in other countries) in favor of a liberal law of abortion is, that the woman's body is her property, and that she (for that reason) has the right to decide in matters concerning her body. It is understandable that the rhetoric of property is useful when powerful groups use religious fundamentalism or other kind of fundamentalist strategies against the (legal) right to abortion. But according to the argument sketched above, the rhetoric of property is rife with pseudo-arguments. It pretends that there are *ontological* solutions to the ethical challenges concerning abortion, challenges which can be met by sound ethical arguments.

VIII. BODY AND PROPERTY

The limitations of the discourse of property and the necessity to supplement it with a framework for ethical analysis is a main theme in Hegel's *Philosophy of Right*.

Hegel's analysis is highly relevant to understanding recent cultural and ethical discussions in north-European welfare-state democracies. Hegel tried – among many other things – to illuminate the relationship between human freedom and the role of the (welfare) state in meeting the needs of its citizens. His anti-foundationalism and his focusing on the concept of freedom reflects the culture of modern democracies. And he actually tried to highlight the ethical legitimacy of the welfare state and to show that this kind of state is not in contradiction with the basic concept of human freedom, but on the contrary is implied by this very concept.

Hegel does not explicitly refer to or discuss Locke's theory of property rights. But it is evident that Hegel offers an alternative to the Lockean theory. According to Locke "Men, being born, have a right to their preservation, and consequently to meet and drink of such other things as nature affords for their subsistence ... Every man has a property in his own person; this nobody has any right to but himself" ([5], p. 16). Locke presents a naturalist argument for private property ([1], p. 67), whereas Hegel's theory is based upon an analysis of human will and freedom. Humans can put their will into things. External objects do not have any end in themselves, so Hegel can contrast his theory with a Lockean kind of theory: "If emphasis is placed on my needs, then the possession of property appears as a means to their satisfaction, but the true position is that, from the standpoint of freedom, property is the first embodiment of freedom and so is in itself a substantive end" ([2], pp. 40-42).

The self is not, as it is for Locke, constituted in the state of nature "ready to appropriate and internalize the objects it needs for biological subsistence" ([1], p. 68). For Hegel the human self and the external would become reciprocally constituted when objects become the property of human will, i.e. when we embody our will in external objects.

It is a consequence of this view that it is contradictory to speak of a right over one's own life. This would be to treat persons as things or animals. As summed up by Brod, "rather than speaking of rights in our bodies; it would be more appropriate to say that for Hegel we have rights through our bodies ... an assault on one's body is not experienced as damage to one's property but rather as direct injury to one's self as a person. Our bodies are the embodiment of our wills, and it is through theories that the will reaches out to the world" ([1], p. 69).

Locke's theory of property can be read as an attempt to provide a naturalist grounding of the rhetoric of property. If the attempt had been successful we would have had reasons against the use of parts of the body without the person's consent. From Hegel's point of view Locke never did reach the level of property *right* at all. Again Brod puts the point in a very acute way: "Locke is speaking only of legalized possession, which may be backed by the force of the state but lacks a properly ethical, political dimension. Rights of any kind constitute relationships not between persons and things but among persons, that is, among embodied wills" ([1], p. 69).

Hegel gives us an alternative to the Lockean theory of property. The Hegelian approach implies that the concept of property is out of place when talking about the human body. So, questions like, 'Should doctors be allowed to do things with me and my body without getting my consent?' cannot be answered by recourse to the concept of property. It is, according to Hegel, non-sensical to speak of humans having rights in or over themselves and their bodies. This does not of course imply that others (the state or the medical profession) have such rights over our bodies. It only implies that the language of property degenerates into rhetoric when used about our lives and our bodies.

IX. PARTS OF MY BODY AS MY PROPERTY

Even if it does not make sense to talk about my body as my property, it might still make sense to talk about *parts* of my body being my property.

I *am* my body, i.e., my will is embodied in my particular body. This body is necessary for expressing myself in external objects which thereby may become my property. However, this does not imply that any parts of me (one of my kidneys, some blood or tissue) is necessary for expressing myself. So, why not conceive of such *parts* of me as external to my embodied will and as something which might be conceived of as my property? In that case I should be free not only to give away but also to sell such parts of me. And nobody else would have this right.

But if we consider it appropriate to talk about parts of our bodies as our property this will not – if we follow Hegel's way of reasoning – solve the ethical problems which arise in relation to medical treatment and medical research.

Conceiving human beings as persons, i.e., as subjects of property rights is an abstraction, – not in the sense that it is a confused or false image, but in the sense that it is a one-sided image which only focuses on some aspects of humanity and relations between persons in the concrete context of modern society.

According to Hegel, the person has a right to life, but it appears only in a very *limited sense of a right to life*: one has the right not to be wrongfully deprived of life through force.

But we can of course be deprived of life in many other ways than by force exercised by others. Lack of food, lack of medical treatment, and a number of other conditions may quite literally destroy my life.

Hegel is not complaining that our idea of abstract rights is the result of an abstraction from these concrete human conditions. Recognizing the abstraction, however, leads us to understand the limitations of abstract rights. Hegel illustrates this clearly by offering several examples in *The Philosophy of Right*. If life, for instance, can be prolonged by stealing a loaf of bread, then someone's property is thereby violated. But, according to Hegel, it would be wrong to consider this action a common theft ([2], § 127R).

It should be clear how Hegel's mode of reasoning is relevant to recent discussions in medical ethics, for example, in relation to the point made by the Danish clergyman: that we do not condone doctors who steal personal belongings of the deceased but (at least until some years ago) we would allow doctors to steal the dead body in its entirety. Hegel might argue that autopsy without consent is not *common theft* (or that the use of organs from the dead for transplantation is not common theft); they can be used as a means for ensuring the personal existence of other persons.

Using Hegel's perspective, our conclusion would be that we cannot analyze our ethical problems about removal of organs, autopsy, etc. simply by reference to the principle of abstract rights. The ethical problems arise in our assessing problems faced by concrete persons in complex, concrete relationships which have been approached in accordance with the principle of abstract rights.

Hegel also tells us why abstract rights are insufficient for analyzing and accounting for ethical problems in our concrete social life. In adopting the standpoint of abstract right we abstract from particular, concrete human *needs*. This eventually implies that we do not have an abstract right to our means of subsistence.

X. A SOCIETY MODELLED ON THE PRINCIPLE OF ABSTRACT RIGHT?

So, appeals to property, then, cannot solve the problems of bioethics, either. It becomes merely ideological rhetoric if we do not recognize that abstract rights are not self-generating. They presuppose, indeed, a social structure with particular norms and values. *Right – according to the Hegelian approach – presupposes a relational structure of reciprocally recognizing persons*. Such a structure has developed historically. It is an expression of particular habits and customs, what Hegel calls *Sittlichkeit* (ethics). 'Sittlichkeit', however, does not only refer to an entire set of institutions; it also refers to a disposition on the part of individuals toward their social life and "an attitude of harmonious identification with its institutions" ([8], p. 196).

Here we face the limitations of Hegel's theory vis-à-vis ethical problems facing the Danish health care system. There is today no unified ethical standpoint embodied in all social institutions. Contradictory ethical perspectives are embodied in different institutional frameworks even in Danish society.

The institution of the Danish welfare system, especially within the health care system, traditionally have been characterized by professional paternalism. This of course is changing, but due to former medical paternalism many Danes nowadays do not identify with the institutions of the welfare state (i.e., they do not recognize these as *their* institutions). Professionals are being criticized for espousing their own *particular* interests as if these were interests shared by the population.

There are, of course, many other institutions in present-day Denmark than the institutions of the welfare state (the family, the local community, folk high schools). These institutions embody an ethical standpoint as well as do the institutions of the welfare state. Traditionally, communal Danish institutions have been characterized by democratic openness and a respect for different opinions and points of view.

Nowadays this democratic society has begun to succumb to a relativistic ethics, identifying the individual citizen as the only authority in ethical disputes. Aspirations to achieve communal consensus through democratic dialogue seemed to slowly fade away. The Danish discussion concerning brain death criteria illustrates this circumstance. Contrary to what is the case in most other countries, the question of the medical criterion of death was in Denmark not considered to be a question for experts alone. A broad public debate on the question was initiated by the National Ethical Council. From the very beginning it was accepted that anyone could participate in the debate which, in the end, should lead to a decision and even a law in Parliament (if the traditionally criteria should be supplemented). After years of discussion and controversy, the question was decided in the form of a law regulating organ transplantation. In the middle of the 1980s, strong pressure groups within the medical establishment worked to convince politicians and the population that heart transplantation should be available for Danish patients. This would, they argued, require acceptance of brain death criteria (for the donor must be legally dead *prior* to the removal of organs).

The public debate was brought to an end in 1990: a law was established to permit heart transplants in Denmark. According to this law, death can be established *either* by the cessation of breathing, heartbeat, circulation and pulsations, or by the irreversible loss of *all* brain functions. Organs may be removed from deceased persons (declared legally dead) who have given their permission (or, if relatives have given permission and nothing indicates that the deceased person would have objected to being an organ donor). This solution is an expression of a democratic society valuing personal autonomy above everything else. But it may equally be viewed as a society having recourse to a narrow, relativistic conception of personal autonomy and principles of abstract right, because it has abandoned common norms and values, its own ethics.[1]

The rationale behind the public debate initiated by the National Ethical Council concerning the criterion of death was that this was not just a matter for medical experts but should be discussed and decided within a cultural and socio-ethical context; or, to put it in Hegelian terms, to be determined within the framework of Danish ethics (*Sittlichkeit*).

There has been, however, no serious discussion concerning the transplantation issue. The experts had their own way. The only ethical aspect in the justification of the law was recourse to abstract right: Danish citizens should have the right to all available life-saving treatments. There has thus been no serious discussion about priorities in health care and the important ethical dilemmas associated with this problem.

The ethical debate initiated by the Danish Parliament and the National Ethical Council became reduced (perhaps due to the lack of a shared ethical framework) to an articulation of individual opinions, regularly summarized in opinion polls. Politicians then tried to 'balance' the data from the polls with the direct influence of specific pressure groups. The right to treatment was considered along with the right to donate or to refuse to donate organs. In such a climate the meager or minimalist ethics of abstract right, that ignores concrete human and social relations, will end, as it begun, in a confused metaphysics of the human body.

Department of Philosophy
University of Aarhus
Denmark

NOTE

[1] For an account of different conceptions of autonomy, see [4].

BIBLIOGRAPHY

1. Brod, H.: 1992, *Hegel's philosophy of politics*. Westview Press, San Francisco.
2. Hegel, G.: 1942, *Philosophy of Right* (Translated by T.M. Knox). Clarendon Press, Oxford, UK.
3. Jensen, U.J.: 1987, *Practice and Progress. A Theory for the Modern Health Care System*. Blackwell Scientific Publication, Oxford.
4. Jensen, U.J. and Mooney, G.: 1990, *Changing Values in Medical and Health Care Decision Making. Changing Values: Autonomy and Paternalism in Medicine and Health Care*. John Wiley & Sons, Chichester, UK.
5. Locke, J.: 1970, *Two Treatises of Government*. Cambridge University Press, London.
6. Macpherson C.B. (ed.): 1978, *Property*. Basil Blackwell, Oxford.
7. Putnam, H.: 1981, *Reason, True and History*. Cambridge University Press, Cambridge.
8. Wood, A.L.: 1990, *Hegel's Ethical Thought*. Cambridge University Press, Cambridge.

FRANZ J. ILLHARDT

OWNERSHIP OF THE HUMAN BODY: DEONTOLOGICAL APPROACHES

I. INTRODUCTION

What the body is and to whom it belongs are questions to which it is difficult to find satisfactory answers. Not to have given considerable thought to these questions however, cannot be reconciled with 'good ethical practice'. It would seem that much depends on the answers to these questions. But if one cannot satisfactorily say what the body is, and to whom it belongs, it would be better not to pretend that satisfactory answers existed. Indeed we would be best served by taking up the work of gaining an orientation with respect to these questions. I understand deontology as something that does not relate only to prohibitives such as 'never to do this'. To set prohibitives is the last step of a process considering the intrinsic challenge of an action we plan, discuss, want to forbid, to recommend etc.

The transplantation of kidneys, hearts, livers and bone marrow has become common practice in the field of transplant medicine. Operations undertaken involve reliable diagnoses, internationally-regulated access to organs, standardized technical procedures, and methods of risk-minimalization. Combined heart-and-lung transplant, and transplant of the pancreas on the other hand, belong to the field of experimental medicine. The question of death arises at the moment organs are to be removed from a donor.

Here the special problem of utilizing the organs of anencephalic new-borns can't be discussed. A significant danger exists in the tendency to take an instrumentalist approach to the body, thereby distancing oneself from the issue of body ownership and the question of whether the body's worth consists exclusively in its possible use.

Two separate approaches, both attempting to overcome the scarcity of organs and at the same time avoid moral problems, have received much recent discussion. The first is the production of artificial organs, and the second involves the use of non-human donors. The artificial-organs

H.A.M.J. ten Have and J.V.M. Welie (eds.), Ownership of the Human Body, 187–206.

approach fails primarily in making devices sufficiently complex so as to integrate compatibly and effectively with the complexity of the body, while the use of non-human organs, a monkey's heart, e.g., currently presents overwhelming challenges to our understanding of immunosuppression.

II. A SKETCH OF THE IMPORTANT ETHICAL PROBLEMS

There are today two main ethical issues in the field of transplant medicine. The first concerns the task for the doctor and patient of arriving at a responsible assessment of the risks involved and, having done that, moving on to any remaining factors in the decision of whether or not to transplant. The second issue concerns the need for sensitivity on the part of the doctor in acquiring the consent of the donor. This is more problematic than might appear. Ideally, a prospective donor's decision should be made in the absence of coercion. But consider the following typical formulation of a patient's medical condition: "If you do not donate your ..., a specific individual will lose his only hope of continued life" ([35], p. 27). Is there something inherently coercive about this approach? Is the physician required to assist the prospective donor in coming to terms with the moral pressures at hand? Basic issues of this kind shall be outlined.

Regulation: By Consent or by Countermand

In Europe, two basic ways have been proposed to regulate organ donation. The first is by explicit consent, and the second by explicit countermand. In the cases of regulation by consent [37] an organ can be removed if the donor, or his legal representative, gives consent. If no donor certificate can be presented, the removal of organs following the consent of the proxies, e.g., is problematic since the necessary discussion must take place in a setting of tremendous stress, and even trauma or shock. Organs can be preserved only for a limited amount of time, making it of utmost importance to obtain consent as quickly after the accident as possible. Time constraints often preclude approaching the situation humanely.

On the other hand, some countries, Austria for instance, attempting to avoid the problem of organ scarcity, favor regulating organ donation by countermand [21]. According to this regulation the consent of the newly-deceased will be presumed in the absence of a certificate expressly stating the subject's disapproval. One effect of the regulation is that it removes such anguished decision-making from the hands of concerned proxies.

Both approaches respect the autonomy of the individual to decide for himself whether or not to donate his organs. It is common knowledge that many people would be prepared to donate their organs, but as a matter of course neglect to obtain a certificate, or fail to renew it. This may indicate an advantage to regulation by countermand. The form of indicating willingness is an expression of freedom and self-determination, the duty of formal documentation, a reasonable protection from coercion. Nonetheless, the danger is significant that one's views concerning the ownership of organs corresponds exactly to one's views concerning the ownership of things. This would involve an unfortunate individualistic and possessive interpretation of the concepts 'having-a-body' and 'being-a-body.' Neither the questions of what 'having-a-body' and 'being-a-body' mean, nor the question of whether one possesses a body can be individually and privately answered.

Thus I think that regulation by countermand may be more suitable. 'Having-a-body' and 'being-a-body' have to do with the social fabric of our existence, and take on meaning only through the realities of contact and sharing, and remain concealed by an anti-social mentality. Nonetheless, the fear of being declared dead prematurely, or of losing one's identity through organ removal, however ungrounded these may be, are to be respected because they belong as elements to important discourse.

Feasibility / Usefulness

A question for doctors and patients alike concerns the feasibility of the procedure. This is perhaps especially so in the case of high risk groups, liver transplant for former alcoholics, e.g., or kidney transplant for very old patients. From a deontological standpoint, too, it is important to take the variables of economics, feasibility, and usefulness into consideration. On a deontological analysis, however, a patient is not to be compared in any way to the cost of the treatment. That would be logically impossible and inhumane. Nonetheless, it is more and more the case that the method of treatment is determined by a cost-benefit analysis, cashed out in terms of the construct QALYs (Quality Adjusted Life Years). Such an analysis is at best highly subjective and pseudo-economic since, quite clearly, not all of the variables are quantifiable, and thus, with no stable point of reference, are certainly incommensurable. For the proponents of racist ideologies it is a matter of assessing the worth of a life from an external perspective, and not from the patient's perspective [4].

As the construct 'quality-of-life' is seen to play an ever more important role in decisionmaking the suspicion suggests itself that it is really a matter of certain preconceptions about others, e.g., the elderly [14]. As long as the question of treatment feasibility and usefulness concerns only medical assessments the situation remains unproblematic. However, the question can lead to further considerations about whether a transplant is worth it, in the case of this patient or that patient. Such considerations cease to concern the feasibility and usefulness of the treatment exclusively, but move outward and factor in the usefulness to the patient. Kant once wrote that people, unlike things, do not have "a worth, but a dignity" ([17], p. 434). On Kant's view people cannot be appraised or have their value compared. Cost-benefit analyses in the field of transplant medicine are to this extent inappropriate.

Quality-Of-Life and Self-Concept

Organ transplants become necessary when the original organ is so irreversibly damaged that transplantation offers the only hope for an improvement in quality-of-life, or even of survival. A life cannot simply be understood as a biological datum, rather it must be seen from the perspective of the person whose life it is. In addition to issues having to do with medical assessment there are the issues of how the recipient of the life-saving healthy organ understands his life; what he expects from his 'new' life; how he wants to determine it; how he expects or wants to come to terms with his 'new' body; and how he wants to feel in his body (many later feel as if 'reborn').[1] Such is the array of issues concerning quality-of-life and self-concept. With respect to the issue of survival neither of these factors can be dismissed as unimportant. Quality-of-life is something which plays a *de facto* role in the physician's deliberations, although the concept itself is vague and often ideologically laden [15]. Nonetheless, the quality-of-life factor inevitably accompanies the various deliberations – the problems of 'inner-sectoral' and 'intra-sectoral' allocation ([30], p. 42), to name a particular example – but also the considerations the hospital staff make as to which kind and how many transplants can be taken up in the budget, or of a department as to which patients should be accepted for transplant, or of a station team as to which level of urgency a transplant candidate should be assessed. It is unavoidable (even if undesirable) that criteria such as degree of complication, mobility, capacity for enjoyment, social integration, and chances for rehabilitation get factored into the overall deliberations. It is not only a problem of how quality-of-life issues can be addressed, but also the extent to which the decisionmakers are aware of the implications.

The self-concept of the transplant candidate also comes into play [3]. A new organ will offer him new possibilities for living in and with his body, and also a chance, perhaps, to recognize and to correct the overly risky features of his previous lifestyle, and to develop a new appreciation of living through-his-body and not at the cost of his body. Such considerations will invariably arise as soon as one recognizes the connection between health and lifestyle. It is important however to remind ourselves that these issues do not constitute a criterion – as a criterion they are far too nebulous – by which the question of whether to transplant can be decided. Quality-of-life and self-concept are issues for the physician to discuss with the patient, should transplantation be deemed feasible.[2]

III. ORGANS: SCARCE MEDICAL RESOURCES

Transplant medicine involves the scarce supply of organs. The more scarce the organ the more restrictive the regulations for its distribution will be. Just what procedure of distribution can be considered remains problematic. Procedures of distributive justice, such as deciding randomly, by lottery, or on the basis of a first-come-first-served principle have received much recent attention. These discussions can be seen as moot, however, since the task at hand is to reconcile the tension between a 'natural' and a 'social' lottery ([6], p. 343). A prerequisite for establishing this, however, would be an acceptable 'view of the good life'. What results is the "ideal of a neutral bureaucracy" ([6], p. 341). Precisely this bureaucratic ideal of administering scarcity doesn't fit the human concept.

In Europe, the distribution of organs is regulated by a central office which has at its disposal all the relevant medical data – concerning, for instance, the compatibility of donor and recipient, and a candidate's up-to-date urgency assessment – needed to make optimal decisions. The procedure involved is dependent on the medical specifications of organ matching, and thereby precludes arbitrary or partial decisionmaking. The judgements which determine a patient's placement on the urgency list, however, can be subjective.

A Critique of the Repair Mentality

The successes of transplant medicine seem to demonstrate that a human organ has a functional value, and that the meaning of an organ beyond that is pure speculation. But those who recognize only the functional value of an organ fail to appreciate the psychic crises faced by transplant candidates. To accept a value beyond this strictly functional value does not mean one is opposed *per se* to transplant surgery. Rather, it means to be opposed to regarding organs as no more than functioning replacement parts, and the role of the physician as that of the repairman.

Is this to overdraw the case? An organ is transplanted to insure proper bodily function and to make possible an improvement in the quality-of-life of the patient. While that much is certainly correct, we shouldn't lose sight of the fact that the concept 'quality-of-life' is not only highly imprecise from a medical point of view, but highly imprecise from any point of view, unless its content is filled in by the patient him- or herself. Quality-of-life cannot be a reason to transplant.[3] Much more reason to transplant is helping the person to balance out his/her commitments and personal life. Often we hear 'I can't do otherwise', when we should hear 'I don't want to do otherwise' – an expression of preference. Why does someone want something? The physician should take this question into account.

Employees working at dialysis units with patients for whom a kidney transplant is foreseen often show signs of extreme withdrawal ([27], p. 582). Studies show that not only patients, but also the members of the treatment team need help in coping with the stress and anxiety associated with the fear of organ rejection. In some cases this need to cope with the anxiety, usually present before and around the date of surgery, has been seen to persist well beyond surgery, sometimes for a lifetime. Body-image problems, as illustrated by the phenomenon of post-operative phantom pain, confirm that one's sensory apparatus too plays a role in coming to terms with the identity and significance of the new organ [3]. This makes clear that the problem of what to do, so as to minimize the danger of rejection, begins with the new organ itself and works its way outward to the personal organization and balance of one's lifestyle. For this reason, psychiatric liaison services providing crisis intervention as well as long-term assistance for the patients ([27], p. 589; [9]) are now often attached to transplant clinics.

Public Obligation

In our highly complex world, individualistic interpretations of human interests are inappropriate. For instance, while it is desirable to protect man's right of privacy against computerized data collection something beneficial, like a cancer registry, would be unthinkable without free access to data. To be sure, there is no opposition between critical data protection and epidemiology, but a balance between both needs will not result from following an individualistic model. The concept 'a citizen's public obligation' [29] should take account of both hazards: that of the totally data-assessed human, and the human who, although living together with others, refuses to share interests with others. Problems in the area of transplant medicine cannot be solved according to the individualist's model. A sense of social responsibility should prevent an individual from seeing his organs as something, like other assets, to be hidden away for safe-keeping.

If addressing the issue of 'being' is unavoidable in this context; it will be helpful to keep in mind that 'being' in the philosophical tradition stands for everything that exists, for things and creatures together. It does not follow from this however, that one has a duty to give up one's own claims, and see oneself as nothing more than a part of the community. To put the question straightforwardly, does a person have a duty, given the scarcity, to put his organs at the disposal of the medical community? 'No.' But perhaps a person has no right to retain his organs for himself. It would have to be the case that he has important reasons for his opposition. This does however, invite the question of whether one can expect to receive an organ in conditions of scarcity, never having indicated a readiness to be a donor. Perhaps an organ should "better be seen as a kind of 'club benefit' with contractual inclusion and exclusion options" [19]. This emphasis on reciprocity seems reasonable.

IV. WHY TAKE A DEONTOLOGICAL APPROACH?

The connection between the need to transplant organs and the attempt to establish satisfactory criteria for what counts as brain death can be seen to lead in a dangerous, perhaps dehumanizing, direction. A person is not simply dead, but is 'declared' dead – a distinction emphasized especially by H. Jonas [16]. Between 'being dead' and 'being deemed dead' there is a confusing gap. For instance, the Harvard Medical School established brain criteria for declaring death in 1968, citing as a reason: The gain of organs which, according to the old criteria (cessation of heart and lung function) would not be available for the purposes of transplant. The fear that someone could not 'really' be dead before his organs get taken out, however unsubstantiated, will remain intractable if the criteria of 'being dead' and 'being declared dead' get confused. In its essence this is a problem of the identity of the body, reflected in the questions: What does it mean to have a body? To be a body? and To whom does the body belong?

The widely accepted consequentialist moral theory is especially ill-suited to addressing these questions, as Martyn Evans has shown (in this volume). According to this theory what is important are the consequences of treatment for those involved. One measures the consequences simply in terms of their social implications, that is, to what extent the treatment can be expected to bring about the greatest good for the maximum number of people. Not taken into account, however, is the issue of whether the treatment represents an appropriate understanding of the body. One of the founders of utilitarianism, J.S. Mill, seems to have been aware of a certain dilemma behind the utilitarian calculation when he posed the question of whether it would be better "to be a Socrates dissatisfied than a fool satisfied" ([29], p. 9), or, less sarcastically, if an action is to have good consequences in order to be called good, or whether it is to be in keeping with a duty towards the object of the action. Appropriateness or suitability is also a matter of the nature of the person who acts.

The other moral theory I shall consider is 'deontological' (from the Greek *deon* = duty) theory because it arises out of the duty which the agent accepts to respect his own moral ends and those of other persons who might figure in as the objects of his action. One of the main proponents of a deontological moral theory, I. Kant, emphasizes that moral content is to be found not in the consequences of an action but in the goodness behind an action: "All of the sciences have some kind of practical part consisting of certain means for the purpose of making a certain end possible ... A prescription required by a physician in order to completely cure a patient and one required by a poisoner in order to make sure of killing him are of equal value so far as each serves to bring about its purpose perfectly" ([17], p. 415).

From this it follows that medical treatment cannot be judged only in accordance with the rules which form its basis. The consequences alone can also not be used as a measure. Otherwise we would have to concede, absurdly, that the mixer of the poison committed an act comparable to that of the doctor. An action must also be judged in accordance with whether the agent acts in such a way as to connect his own requirement to treat well with the claim of his fellows to be treated well. With respect to transplant medicine this means the balancing of professional duties – of the value of medical experience, for instance, or the value of the long and short term social effects – will not help when the terrain is unfamiliar [2]. Metaphorically speaking, in a landscape which has just been opened up, and in which there are useful directional signs, it is often good to weigh the one and the other direction, to calculate detours, to find out in advance whether the route is comfortable, quick, direct, and picturesque, etc. However, when the existing directional signs themselves are to be checked for accuracy one needs a set of coordinates in order to determine if a particular route is appropriate. In the routine of transplant medicine a balancing out of the likely consequences is certainly advisable. However, an assessment of this moral argumentation in everyday life needs fundamental criteria and deliberation. Balancing and checking of details comes later.

In modern medicine it doesn't make sense to maintain a polarity between consequentialist and deontological moral theories. The balancing out of consequences should involve an assessment of whether or not a planned treatment, or one just carried out, respects the identity of the patient and the general coherence of his ends, so that they are not subverted in favor of a positivist calculation.[4] Likewise a deontological theory will have to take the expected consequences of a procedure into consideration in order not to be blind to the concrete conditions of interaction. A useful formula for this approach requires that we consider the meaning of treatment beyond its social usefulness, that is, that we consider what 'concerns us absolutely' (P. Tillich). If this question does not get asked the distinction between physician and poison mixer will begin to dissolve.

If deontological theories have come to have a bad reputation it is likely on account of the misinterpretation that such theories require a direct, positivist application of a rule or principle to a moral situation. A principle, that the autonomy of an individual is to be respected, for instance, is supposed to offer moral guidance to the physician in the case of transplanting an organ. There is a tension between the levels of deliberation. The deontological principle is couched in highly general or highly abstract terms while the questions facing the physician appear to be concrete. Highly general answers to concrete questions tend to sound wishy-washy. Here it will be helpful to keep two things in mind: (1) The discrepancy in the levels of generality is not only a problem for deontological ethics, but rather a problem for any moral theory which is based on rules and which seeks to derive practical guidance from general principles. Thus the deontological principle must get as close to the concrete problem as possible [9]. (2) It is not clear that an ethical theory is to issue answers in the sense of instructions for action. What is helpful are not answers *per se* but problem analyses which point out how one might productively approach a problem. Deontological moral theories are especially well-suited to bringing into the foreground the basic elements of ethical experience which in the normal course of procedures tend to get overlooked [10].

One purpose of taking a deontological approach in medical ethics lies in pointing out treatments which are bad because they are intrinsically bad. However, the appropriateness of such an approach is not to be found in its capacity to intervene in medical practice so that such treatments do not get performed. Precisely with respect to ethical deliberations in medicine it is important to emphasize that deontology does not constitute a decision procedure, it is rather a theory of orientation and justification. Even the most rigorous deontological precepts, for instance, those of Catholic doctrine, were thought of as overridable, given certain circumstances.[5] It is never permissible to kill; however, the examples of self defense, the death penalty, and the just war represent concrete situations where the general rule doesn't yield a suitable answer.

Furthermore there are many different kinds of deontological theory, which by no means display the often charged rigidity. What is more, deontological considerations can be brought to bear in the case of a single action, or in the case of the rules and principles with which one seeks to assess the merits of such actions. Deontological considerations track "right-making characteristics" ([1], p. 36) which occupy a legitimate place in our everyday moral disputes, for instance, as moral intuition, as common sense, as appeal to human rights or a kind of moral contract, but also as moral principles, or in the appeal to human values and needs ([1], p. 37-41).

One function of ethics is the inspection of existing mores or, one might say, ethos critique. The deontological approach is especially well-suited to this task because it makes it possible to leave the framework of observable action. It concerns itself with the elements which lie at the basis of action and which determine the goodness of an action. A consequentialist approach must remain within the framework of the observable sequence of actions in that it focuses on consequences, and thus can take up a critique of the action and its basic orientation only to a certain extent. Merleau-Ponty describes deliberations which seek a moral optimization of technique (but which are incapable or indisposed to make broader assessments), as "the 'operative' thought' in which 'human creations [creations, like transplantation] are derived from a natural processing of information which is, however, itself designed on the model of a machine". ([24], p. 14). Consequentialist thinking based on the means-ends ratio cannot raise fundamental criticism, for instance, of an action whose moral underpinnings concern what is deepest and most important about us, and which often lie on the other side of social accommodation. An examination of the consequences alone will not bring us any further in resolving moral problems in modern medicine. Fundamental approaches are required. Deontological considerations can highlight our basic moral features, and thus certain borders, which are not to be crossed. Such borders become clear when one reflects on what our embodied state means, that is, for what purpose it is called upon, and also when one considers the context of the concept of 'belonging'.

V. THE 'A PRIORI' OF THE BODY

It would be a travesty to try to distinguish body from soul in a human being. The modern era – ever since the physician and philosopher Descartes – has labored under the conceptual confusion of such 'separateness', and this has lead to the view that bodily processes can been delineated and treated as something other than the soul. What gives the lie to this view is that every kind of reduction contradicts the experience of being-a-body and the experience of complexity: complexity of the objects with which we are concerned, complexity of action in general and in the field of medicine in particular.

A human being and his corporeal existence is especially compellingly seen in his relationship to God. It is widely assumed that in the biblical order of creation man does not belong to himself but to God, and that this God determines the ends of man. But this manner of thinking doesn't take into account the creator-creature relationship, and overlooks the fact that creation theology is also responsible for introducing a special view of being a creature. Creatures are not simply things belonging exclusively to the order of nature, but rather have a 'piece' of the reality of the creator. They are in important respects similar to him, they have a God-like creative potential. On the other hand they are to be understood in light of their existence among other creatures, sharing their reality with them. The relationship between creator, creature, and fellow creatures cannot be understood as one of dominance and subservience, but rather as one in which the creature owes his creator a certain debt or responsibility. The human being stands to his fellow creatures in a relationship of trusteeship. He cannot dispose of them as he will. Rather he must recognize their autonomy. He too has autonomy. He is not only something made by God, but rather capable of deciding and acting in the sense of creation.

Contrary to certain Platonic views this moral agent is not a composition of body and soul, whereby the soul (or mind) is responsible for thought and purposeful ends while the body (matter) is responsible for their material realization. In the case of biblical anthropology the human being is a single unit. The uniqueness of humans is not that God first created the body to which he then added the soul. For these essential reflections the issue of the 'substance' of body and mind is not problematic. What is questionable is what makes humans human. According to biblical metaphor it is *nefesch* (breath), and it implies the empowerment of the body to be a human body ([36], p. 35-37).

This creation theology is the starting point for further deliberations about human corporeality. Man made the world subservient and understood this word 'subservient' (in keeping with the ethos of the Greco-Roman slave-holding society) as standing at ones disposal and putting oneself at the disposal of another. Two unsuspected experiences weakened this system. The first was the possibility of the absolute subduing of nature, that is, of the world of organic matter. The second was the possibility of atheism. A new conceptual scheme was required to make room for these new ideas. One such new thought comes from the history of medicine. Human beings came to be seen less and less as an impenetrable mystery. Increasingly it was recognized that medicine has a duty to explore the micro and macro world inside and outside of the human being. This empiricism began to collide with anthropological experience. Focusing on single organs, tissue, or functions, was not seen as self-evident as it is today. On the contrary, it was not regarded as a viable approach.

This exploration of the human being was experienced as a threat to the understanding, prevalent at the time, that human beings were essentially mysterious. The discovery of the circulation of the blood by Harvey in 1628 can be considered an example of this conflict [11]. He calculated precisely the amount of blood in the body, and described the human heart as the central function of the human being, thus transferring the principle of life sustainment inward, into the human body itself.

Until that time, according to the principle of life sustainment, the sustainment of life was thought to inhere in the king. What the sun was for the planets, the king was thought to be for the people. Harvey described the heart as the sun of the body, making a point of comparing it to the function of the king, and with that he dethroned the external guarantor of life, the king, and transformed the human into a self-sustaining being. To that extent it was not only a matter of an anatomical understanding of the human, but also an anthropological understanding ([34], p. 9), in that a human being could no longer be seen as within a hierarchy. There was a danger in the political consequences of Harvey's understanding of the human being for religion and the monarchist constitution of the land [31]. The reference point of the human being no longer lay in a God-given social order, but in the human being himself, who is himself responsible for social order.

These ideas crystallized in the concept of 'autonomy' which on one reading emphasizes the social dimension of human existence. Accordingly human existence came to be thought of as inhering in a kind of 'relatedness'. We can credit Kierkegaard for having brought this idea to the fore in opposition to the prevalent idealist tradition in philosophy. This surpassing of earlier traditions was made possible through careful observations of bodily processes. One important conclusion to draw from this is that the body is not simply something organized around an immaterial mind. It is 'in' the body and nowhere else that the meaning of the human being is to be found.

The French philosopher Merleau-Ponty uses an example to highlight this point. For the painter a body is an object of human perception. With the same logic Merleau-Ponty could also have spoken from the perspective of a transplant surgeon who removes an organ from one body, grafts it into another, and observes how the one body comes to terms with the other. However, he formulated a simple example of how the right hand experiences the objectivity of the left hand. An extraordinary event takes place. The bodies perceive one another and are something other than what they were before. They become animate ([25], p. 52). Merleau-Ponty infers the following: (1) There is an essential relationship between body and consciousness such that the body is never – even throughout transplant surgery – just a body, but rather a perceiving entity, that is to say animate. (2) Every body receives its specificity and becomes animate through the perception of another ([25], p. 52f).

Are these considerations too abstract to be thought of as helpful in addressing the problems of the various fields of medicine, influenced as they are from the hard sciences? To be sure this body-holism philosophy hardly seems convincing or useful with respect to the problem of immuno-suppression. But just as clearly it is worth mentioning that precisely in the case of immuno-suppression an event is there to observe in which one body recognizes another as foreign and undertakes to 'reject' it. Is it possible that the immune system has the significance and function of protecting identity? From where do these amino acids get their apparent 'wisdom' [33]? They appear to be acting on behalf of the entire body, although they constitute only a part of it. The body is more than its functions. This can be the sense of the 'principle of totality' [12] received from the theological tradition. In the presented interpretation the principle contains a warning against understanding the body as a collection of components rather than a guide to action.

In medicine bodies of all kinds are exhaustively understood. But does this mean that there is also an understanding of 'being a body,' and is this taken into consideration in deciding at what point to replace an organ, or at what point the integrity of the body must be respected? Medicine is today grappling with the concept of quality-of-life. Medicine seems instructed to do so, and the result is that the need to address the deeper issues of the body is getting ignored in favor of a more superficial discussion where little admits of quantification.

Everything in medicine concerns an ensouled, perceiving, and perceived human body. One seeks to avert a disaster by replacing a diseased organ with a healthy one. Success has given a certain credibility to medicine which should not be denied. Only occasionally a problem appears for which medicine offers 'new years to a life, but not new life to the years' (H. Schipperges), that is, without asking if perhaps a good death would not have been the better alternative. It belongs to the "tragedies of modern medicine that physicians now have the possibility of saving a life in cases where the economic, personal and affective capacities of society stand in opposition" ([32], p. 230]. The same problem exists when it is a matter of developing synthetic organs such as the Jarvick-7-heart. The question of whether this approach is humane cannot be determined by a simple straightforward appeal to the more or less beneficent intentions and consequences. An important measuring stick is whether or not the new organ makes possible, not only a few enjoyable days, but a satisfying extension to ones life. One is forced to take a long look at the cynical symbolism of it all. There is the artificial heart's switch with which the patient can bring an end to the activity of the heart and thereby to his life, should it become unendurable. Is this, in the end, the most human aspect of it all, because it shows an understanding of the problems of embodied existence?

VI. THE MANY-LAYERED CONCEPT 'TO BELONG'

One of the first things to clarify for any aspect of transplant medicine concerns the linguistic connotations of the verb 'to belong'. Marilyn Monroe might still sing, untouched by psychological and ethical qualms, 'My heart belongs to daddy.' Are we to take this literally? Naturally a daughter doesn't thereby declare her father's exclusive right to obtain her heart for the purpose of surgery. Nor is any right of possession handed over. Of course, what the daughter means to express is only how dear her father is to her, and the heart is simply a metaphor for this relationship. Looked at purely statistically, it is by all means possible that fathers become violent when their daughters no longer want to be daddy's little girl or, what has been much discussed of late, that children develop aggressions against their parents as they grow older.

It is also important to keep in mind that the expression 'to belong' comes from having a right over things, and that, historically speaking, people have often been regarded as things which stood at one's disposal. Issues like the abuse of women and children, and, of late, that of the elderly, demonstrate the archaic notion that people are possessed as things, that people take themselves to have a right, not only in relation to other people, but also 'over' others, and that one is therefore entitled to mistreat others as if they were things. The *pater familias* to which the family belonged and which had an absolute right of disposal over each of the family members is a legal relic from antiquity. Family law drew the demarcation between the sense of 'belonging to' and 'having dominion over' even more narrowly, but didn't repudiate the virtual right of possession of one person over another. This law has been modernized in most countries only in the last several decades. In other words the metaphors of antiquity reach as far as the present.

A famous German mystic of the middle ages was Meister Eckhart [23]. Out of his teachings an 'ethics of release' (W. Weischedl) was formulated, according to which the possessive element with respect to people and things was to be eliminated and freedom to create the world and the self was to be taught. Two important things follow from this. First, people and things resist being 'possessed'. They have their own uniqueness which people can experience, but over which they are not to exercise control. Paradoxically formulated, one can have things and people only in so far as one permits their freedom – hence the name 'ethics of release.' Second, only he who permits the freedom of the other can discover the treasure of the other. He who has something tends to see only its usefulness. To have things and people and to place value in having them is dangerous because this 'mode of having' (E. Fromm) and the model 'I can only live if I have ...' clouds the consciousness and prevents a better appreciation of the unconditionality of existence.

With this in mind it becomes clear what misunderstandings are involved in the possessive demands that turn up in relationships between people, even in relationships of love and friendship, misunderstandings which can turn a human gift into a psycho drama instead of a case of respect and devotion. In the background of the ethics of release it is clear that any kind of desire to possess will destroy the freedom of the other and that of oneself.

In his *Metaphysics of Morals* Kant argues in favor of the freedom to enter into relationships and to refrain from doing so: If there is anything which can really be called good, it would be the will of the person who acts and not the end that the person wills to achieve – which one person may deem worth striving for and the other not. In such a case the goodness of the action would lie outside of the action and outside of the agent. The freedom of the one, supposing the goodness of his will, could not possibly contradict the freedom of the other, supposing also the goodness of his will. Otherwise freedom and a good will would stand in contradiction. From this follows the categorical imperative, that one is never permitted to treat another person as a means only, thereby taking away his freedom to share in the creation of the ends of the action, forcing him simply to endure them. What is a valid end for one must also count as a valid end for the other, even in the case where the actions are presumed to occur simultaneously ([17], pp. 437-440).

Allowing yourself and others to be free is a necessary condition for moral action. Kant does, however, understand the everyday context and conditions of action. He requires that the other never be treated *only* as a means to an end. Action involves interaction. People are always effected by the actions of others, and are always involved in the objectives of others. To them the freedom must remain to consciously take up or refrain from taking up a hand in contributing to the objectives. It is self-evident that the roles should be fundamentally exchangeable. Otherwise the result is heteronomous action. These considerations may sound excessively abstract, but they can serve as a practical guide. In virtually no nation is organ donation mandatory or are transplants performed without the consent of the patient. Nonetheless, the issue of freedom and the rejection of the notion that one can belong to another person cannot be thought of as settled. There are subtle ways of removing a person's freedom.

There is the case of the father who declares himself willing to donate bone marrow to his illegitimate son against his private desires because others expect that of a good father. There is the case of the organ recipient who feels an obligation to prove himself worthy of the victim of the auto accident whose organ he receives. There is the case of the medical student who carries her donor certificate on her person at all times because she thinks that physicians will give her better treatment in cases of sickness, being motivated by the desire to keep her organs fit for the purpose of future transplant. In light of the terminology of this essay can we say that these people belong to themselves, or are they dependent on other people who introduce goals, who have power, and who understand how to yoke other people into realizing these goals?

Belonging to oneself is unavoidable. A particular treatment, however, may ignore a person's freedom. But in that case the treatment cannot be thought of as 'good', even if the consequences of the treatment are 'good'. Subtle forms of not belonging to oneself should not be permitted among the everyday problems in transplant medicine. They dim the absolute worth of freedom. A little bit free is just as nonsensical as a little bit pregnant. In the cases presented above that would mean something like the following. (1) To expect bone marrow donation from the father of the child because it is life-saving, overriding his concerns about his identity, cannot qualify as an adequate ethical solution because it leaves important interests unaccounted for. (2) It cannot be taken for granted that organ recipients form their lives in a certain way out of gratitude towards the deceased donors, even if the artificially adopted life could only be seen as decent and purposeful ([7], pp. 5-10). Decisions concerning ones life plans cannot, in the end, be externally determined. This would contravene the basic claim to freedom. (3) Finally, the student who carries her donor certificate should know she has a right to appropriate medical attention, and need not resort to such self-serving strategies.

Center for Ethics and Law in Medicine
University Clinic Freiburg
Germany

NOTES

1 In the psychology of transplant surgery this is known as the "rebirth phenomenon" ([4], p. 221).

2 I will discuss this in 'A Critique of the Repair Mentality' as a problem of counseling.

3 At an Ethics Consultation Service (directed by George Kanoti S.T.D.) attached to a clinic in Cleveland, Ohio, it is standard procedure to help transplant candidates work through such life issues.

4 G.H. Mead finds a lack, both in utilitarian and deontological moral theories, of an answer to the question of what 'morality' is really supposed to be ([23], p. 381-389). As an answer he proposes 'identity'. It is in this moral-philosophical sense that I use the word.

5 Consider, for instance, the variations of probabilism and tutiorism in Christian theology.

BIBLIOGRAPHY

1. Beauchamp, T.L. & Childress, J.F.: 1989, *Principles of Biomedical Ethics*, 3rd ed.. Oxford University Press, New York, Oxford, UK.
2. Bien, G.: 1981, *Aristotelische Ethik und Kantische Moraltheologie*, Freiburger Universitätsblätter (No. 73), 57-74.
3. Castelnuovo-Tedesco, P.: 1981, 'Transplantation: Psychological Implications of Changes in Body Image', in N.B. Levy (ed.), *Psychonephrology. Psychological Factors in Hemodialysis and Transplantation*. Plenum Press, NY, pp. 219-226.
4. Cohen, C.: 1983, ''Quality-of-life' and the Analogy with the Nazis', *Journal of Medicine and Philosophy* **8**, 113-135.
5. Daniels, N.: 1987, 'The Ideal Advocate and Limited Resources', *Theoretical Medicine* **8**, 69-80.
6. Engelhardt, Jr. H.T.: 1987, 'Shattuck Lecture: Allocating Scarce Medical Resources and the Availability of Organ Transplantation', in D.H. Cowan *et al.* (eds.), *Human Organ Transplantation. Societal, Medical-Legal, Regulatory, and Reimbursement Issues*. Health Administration Press, Ann Arbor, MI, pp. 339-353.
7. Fox, R. and Swazey, J.P.: 1978, *The Courage to Fail. A Social View of Organ Transplants and Dialysis*, 2nd ed.. University of Chicago Press, Chicago/London, UK.
8. Freyberger, H.: 1981, 'Consultation-Liaison in a Renal Transplant Unit', in N.B. Levy (ed.), *Psychonephrology. Psychological Factors in Hemodialysis and Transplantation*. Plenum, NY, pp. 255-263.
9. Graber, G.C. & Thomasma, D.C.: 1989, *Theory and Practice in Medical Ethics*. Continuum, NY.
10. Habermas, J.: 1973, 'Philosophische Anthropologie', in J. Habermas, *Kultur und Kritik. Verstreute Aufsätze*. Suhrkamp, Frankfurt, pp. 89-111.
11. Hill, C.: 1964, 'William Harvey and the Idea of Monarchy', *Past & Present* **27**, 54-72.
12. Hughes, G.J.: 1986, 'Totality, Principle of', in J.F. Childress and J. Macquarrie (eds.), *A New Dictionary of Christian Ethics*. SCM Press, London, p. 629.
13. Illhardt, F.J.: 1990, 'Flucht der medizinischen Ethik vor der 'Lebenswelt' des Menschen, *Medizin-Mensch-Gesellschaft* **15**, 266-72.
14. Illhardt, F.J.: 1993, 'Ageism: Vorurteile gegen das Alter', *Zeitschrift für Gerontologie* **26**, 335-338.
15. Illhardt, F.J.: 1993, 'Hermeneutik des Begriffs *Lebensqualität*', *Wiener Medizinische Wochenschrift* **142**, 523-526.
16. Jonas, H.: 1985, 'Gehirntod und menschliche Organbank. Zur pragmatischen Umdefinierung des Todes', in H. Jonas, *Technik, Medizin und Ethik. Zur Praxis des Prinzips Verantwortung*. Insel Verlag, Franfurt am Main, pp. 219-241.
17. Kant, I.: 1968, *Grundlegung der Metaphysik der Sitten* (Kants Werke Bd. 4. Akademie-Textausgabe). Walter de Gruyter, Berlin, pp. 393-463.
18. Kierkegaard, S.: 1849/1964, *Die Krankheit zum Tode* (Philosophisch-theologische Schriften). Olten/Hegener, Köln, pp. 31-177.
19. Kliemt, R.: 1993, ''Gerechtigkeitskriterien' in der Transplantationsmedizin. Eine ordoliberale Perspektive', in E. Nagel, Ch. Fuchs (eds): *Soziale Gerechtigkeit im Gesundheitswesen. Ökonomische, ethische, rechtliche Fragen am Beispiel der Transplantationsmedizin*. Springer, Berlin/Heidelberg, pp. 262-276.

20. Luhmann, N.: 1988, *Paradigm Lost: Über die ethische Reflexion der Moral* (Festvortrag anläßlich der Verleihung des Hegel-Preises der Landeshauptstadt Stuttgart am 23. November 1988 im Neuen Schloß Stuttgart). Enke, Stuttgart.
21. Margreiter, R.: 1992, 'Die Widerspruchslösung zur Regelung von Organentnahmen in Österreich aus der Sicht eines Transplantationschirurgen', *Ethik in der Medizin* **4**, 185-190.
22. Mead, G.H.: 1967, *Mind, Self, and Society. From the standpoint of a social behaviorist* (Suppl. IV: Fragments on Ethics, 379-389). University of Chicago Press, Chicago/London.
23. Meister Eckhart: 1979, 'Von Abgeschiedenheit (about 1300)', in D. Mieth: *Meister Eckhart*. Walter, Olten/Freiburg, pp. 81-98.
24. Merleau-Ponty, M.: 1964, *L'Œil et l'esprit*. Gallimard, Paris (translation used: *Das Auge und sein Geist*. Philosophische Bibliothek Vol. 357). Meiner, Hamburg, pp. 13-44).
25. Merleau-Ponty, M.: 1960, *Le Philosophe et son Ombre*. Gallimard, Paris (translation used: Der Philosoph and sein Schatten. Philosophische Bibliothek Vol. 357). Meiner, Hamburg, pp. 45-67.
26. Merleau-Ponty, M.: 1960, *L'Homme et l'Adversité*. Gallimard, Paris (translation used: Der Mensch und die Widersetzlichkeit der Dinge. Philosophische Bibliothek Vol. 357). Meiner, Hamburg, pp. 115-34.
27. Muthny, F. et al.: 1985, 'Bedeutung psychologischer Forschung in der Versorgung von Dialyse- und Transplantationspatienten', in F. Balck, U. Koch, and H. Speidel (eds.), *Psychonephrologie. Psychische Probleme bei Niereninsuffizienz*. Springer, Berlin/Heidelberg, pp. 579-592.
28. Mill, J.S.: 1980, 'Utilitarianism', in H.B. Acton (ed.), *Utilitarism*. Dent & Sons, London, pp. 1-61.
29. Schäfer, H.: 1992, 'Die Sozialpflicht des Bürgers', in G. Dhom (ed.), *Epidemiologische Forschung und Datenschutz in der Medizin* (Medizinische Forschung Bd. 4). G. Fischer, Stuttgart, pp. 273-294.
30. Schölmerich, P.: 1989, *Gesundheitsökonomie und ärztliches Handeln* (Akademie der Wissenschaften und der Literatur. Abhandlungen der geistes- und sozialwissenschaftlichen Klasse Bd. 14). Steiner, Stuttgart, pp. 35-48.
31. Schouten, J. and Goltz, D.: 1977, 'James Primerose und sein Kampf gegen die Theorie vom Blutkreislauf', *Sudhoffs Archiv* **61**, 331-352.
32. Seidler, E.: 1985, 'Ethische Probleme', in F. Balck, U. Koch and H. Speidel (eds.), *Psychonephrologie. Psychische Probleme bei Niereninsuffizienz*. Springer, Berlin/Heidelberg, pp. 227-234.
33. Sullivan, M.D.: 1990, 'Reconsidering the Wisdom of the Body: An Epistemological Critique of Claude Bernard's Concept of the Internal Environment', *Journal of Medicine and Philosophy* **15**, 493-514.
34. Temkin, O.: 1971, 'The Historiography of Ideas in Medicine', in E. Clarke (ed.), *Modern Methods in the History of Medicine*. Athlone, London, pp. 1-21.
35. Wikler, D.: 1988, 'Ethical and Ideological Assumptions of Organ Substitution Policy', in M.D. Boulder (ed.): *Organ Substitution Technology. Ethical, Legal, and Public Policy Issues*. Westview, London, UK, pp. 22-30.
36. Wolff, H.W.: 1973, *Anthropologie des Altenen Testaments*, 3rd ed.. Kaiser, München.
37. Wolfslast, G. and Smit, H.: 1992, 'Argumente für die Zustimmunglösung zur Regelung von Organentnahmen', *Ethik in der Medizin* **4**, 191-194.

MARTYN EVANS

THE UTILITY OF THE BODY

I. INTRODUCTION

Perhaps it is because modern health care technology has progressed so rapidly, constantly disclosing new challenges both to our moral sensibilities and to our confidence in our own understanding of ourselves, that we sometimes resort to replacing questions of morals (which are generally difficult and disagreeable) with questions of meaning or of definition (which are generally intriguing and less liable to easy resentment). It is my view that this has happened in the case of the question 'Who owns my body?'. Superficially this is a question of meaning or of definition; underneath it lie questions of morals, questions such as 'What may we do?', 'How should we behave?', 'Whose interests make the strongest claims upon us?'. There is no doubt that we make use of our own, and other people's, bodies in many ways. Equally there is no doubt that we make use of our own and other people's property. In deciding what are permissible uses of property, the fact that it *is* property (that is, legitimately belonging to someone or other) is part of the moral assessment. Perhaps as a result of this, there is a temptation to think that puzzles about the use of our own or other people's bodies can be approached in the same way – by settling first of all the question of who *owns* that body.

Yet I believe the strangeness of this way of proceeding is apparent the moment that we state it, since it presumes that people can be understood and regarded in the same kind of way as property. This seems to be both conceptually confused and morally repellent. Nonetheless we remain confronted with the moral challenge of how we ought to obtain and distribute the various body parts and body products which modern medical science can utilize. Although one possible answer to this challenge is to say 'We shouldn't do this at all', I do not propose to pursue it. The precedents for making use of one another's bodies are sufficiently well established in manual labor, for instance, for us to recognize that new manifestations of this challenge (How should we make use of one another?) must be faced in the terms in which they arise.

H.A.M.J. ten Have and J.V.M. Welie (eds.), Ownership of the Human Body, 207–226.

I do, however, want to establish at the outset that when, for example, I talk about the blood that you may need, though it presently runs through me, if I refer to 'my' blood I need not be talking about property at all. Although the fact that the blood does presently run through me is morally important, this importance arises more obviously from our obligations to care for and not to harm one another than it does from any notions of property, title, or rights of disposal.

A second general point I wish to make concerns the meaning of 'utilitarianism'. There is a widespread conception of utilitarianism as meaning little more than a special concern for the consequences of actions or policies. While this concern is certainly true of utilitarianism, it fails to capture what is distinctive about utilitarian views, in particular their linked assumptions that all consequences can in principle be measured, compared, and aggregated together in a common coinage, and that those outcomes can be identified or predicted which will produce the greatest amount of practical benefit (however this be defined). It is with this stricter sense of 'utilitarianism' that I am concerned in this essay. In the more colloquial, and certainly weaker, sense I suppose that all public policies are 'utilitarian', almost by definition: public policies are designed to produce certain kinds of outcome, by which alone their success will be judged. At the same time, the strict sense of 'utilitarianism' provides the kind of theoretical justification that is most likely to feature in any systematic defence of particular public policies. Part of the reason for this, as James Griffin suggests, is simply that the looser idea of maximizing the benefit has an intuitive appeal that seems to coincide with common sense practical morality [4].

With these preliminary observations made, we may now consider in more detail the characteristics and limitations of utilitarian perspectives, and subsequently their application to public policy in the contested areas of body ownership.

II. CHARACTERISTICS AND LIMITATIONS

What a Moral Theory Cannot Tell Us

Utilitarian theories aim at describing what makes an action or a policy morally right. This is not the same as determining what it makes sense to say. I stress this because there are some things that the utilitarian can say about ownership of the human body, and some things that he cannot. He can say that there is (or that there is not) a moral justification for treating the human body in various ways, for instance using it as a source of blood or spare parts for other people, using it as a source of controlled energy in manual labor, using it as the means to provide sexual favors and so on. But he cannot *as a utilitarian* tell us whether it *makes sense* to regard the body as property.

There have, historically, been traditional understandings of the human body whereby some people's bodies are effectively the property of others – as slave, as chattel, even as wife or child. Modern rejections of such notions can take two obvious forms, and the utilitarian can contribute to – or, for that matter, challenge – only one of them. The utilitarian can have something to say about *moral* rejections of the idea of ownership of the human body (for instance, rejections based on an appeal to the value of every individual). But a different form of objection points to the *conceptual* distinction between things which may be possessed, and people who possess them. That is, we might deny that it makes any sense to regard the human body as property – even my own body as my own property. In speaking of 'my body', I am using an expression not of ownership but of identification – as is plain when I speak about 'my father', 'my favorite football team' or 'my employer'. The utilitarian perspective cannot engage with an objection of this conceptual kind, though it may describe what is useful, and for that matter what is inconvenient, about taking such a view.

Again, for anyone who accepts that in some sense the human body *can* be owned, the question could arise in any particular case: 'Who owns this body?'. This would arise most readily perhaps in the case of a deceased, but as a rhetorical question it is sometimes put forward in the context of debates about surrogate motherhood or live organ donation: 'Is my body not my own to do with as I choose?'. Different answers might be offered in different cases; someone might argue that the deceased's body belongs to the surviving relatives, or to the courts, or to the State, or to the legal executor, or to the hospital/mortuary administration, and so on. We are not concerned here so much with whether any of these is (or could be) the right answer, but rather with noticing that the question of ownership is not initially a moral question. It might be a matter of law, for instance, or of custom. Pressure to change law or custom can of course be prompted by moral considerations, and here utilitarian perspectives could be used to claim varying degrees of benefits for different solutions. One could therefore make a utilitarian case for *choosing* some particular answer or other; but that is not the same thing as settling who (if anyone) is as a matter of fact the lawful owner of a particular human body.

It is important to be clear about the limitations on what a utilitarian perspective can show us about body ownership.

Moral 'Arithmetic'

Utilitarian theories characteristically concern the *sum total* of the good and the harmful effects of actions or policies. The distinguishing feature of utilitarian (as distinct from other consequentialist) perspectives is that they regard the various consequences of an action as capable of being *weighed and measured together*, or 'aggregated'. On this, two things must be said.

First we should note that while as individuals we may often give serious thought to the consequences of our actions, we do so on the basis of those consequences which we can foresee as directly affecting the people with whom we come into contact. We do not conceive of those consequences in terms of their general promotion of the good.[1] Partly this reflects the fact that our individual actions tend to be limited in the range and scope of their effects. By contrast, public or social policies are characteristically conceived in terms of their effects on large numbers of people. So both currently and historically utilitarian perspectives tend to concern the morality of large-scale actions and policies which will affect the public good: innovations and changes in state provision and regulation, welfare, criminal and civil law, and so on.[2] This is where the *importance* of 'aggregating' consequences lies.

Second we must consider the *possibility* of aggregating consequences in this way. It is obvious that we can weigh or measure together only those things which are comparable, that is, which are (or can be reduced to) the same sort of thing. So utilitarian perspectives are distinguished by regarding different sorts of consequences – for instance, the number of new houses built, your success in an examination, reductions in the birth rate, my sadness at the loss of a relative, improvements in peri-operative pain control – as at least in principle capable of being weighed in the same scales. This alone has threatened to make such theories fatally implausible.

Modern utilitarians recognize the problem and some have tried to overcome it by talking about adding up the preferences which individuals express concerning various alternative outcomes, assuming that these preferences can be expressed on a scale which everyone uses in the same sort of way. The more plausible attempts at this are those which confine themselves to the most narrow range of consequences, usually in a specific area of application, for instance the idea of 'welfare' – which is already diverse enough, covering as it does such contributory areas as housing, education, health, food, social security, and so forth.[3] I do not think that we can rehearse here the various refinements that have been suggested, let alone the objections to them. What they indicate is, however, important: the utilitarian perspective is at its most plausible when it bears upon fairly close alternative applications of public policy, where the alternative outcomes can be compared for their impact on large, undifferentiated numbers of people. For instance a decision to choose a particular route for a new motorway in preference to an alternative route a few miles further south can be based on a comparison of the noise, disruption, and so forth that it is likely to cause to approximately known numbers of people. A moral justification for causing broadly comparable disruption to fewer people rather than to more people can be described in utilitarian terms. It is much more difficult for the utilitarian to describe the decision to invest in motorways as opposed to diverting resources into energy research, programs of urban renewal, or increased overseas aid.

These problems of 'aggregation' are clearly important when we bring utilitarian perspectives to bear on the uses that may be made of the human body. If we are considering alternative policies or practices which either involve, or restrict, making practical use of the human body, are we comparing alternative outcomes or consequences which are sufficiently similar as to be weighed in the same scales? If not, then we will not be able to aggregate them in the way which utilitarianism prescribes.

How We Determine 'The Good'

A third characteristic of traditional utilitarian perspectives is one which it shares with other theories of the kind called 'consequentialist' (of which utilitarianism is simply the most widely-known example). Such theories assume that the morality of an action or policy is determined *solely* by its consequences: thus no action whose consequences are agreed to be maximally good can be morally wrong, since what is morally right is derived from what is practically good. Thus from a utilitarian perspective, the ends really do justify the means, because there simply are no other kinds of justification. It follows that any moral objections to a course of action approved by a utilitarian calculation must be an objection to utilitarianism itself, since utilitarianism claims to account for all moral judgements.

I once watched, via television, good use being made of the body of an aborted human fetus. In the now-famous experimental operations in Lund, Sweden, fetal brain cells are implanted into the brains of people suffering from Parkinson-like conditions, whose disastrous impairment of physical function is dramatically relieved by the ability of the implanted cells to produce dopamine, the substance which enables controlled and fluent bodily movement. For sufferers from Parkinson's disease, this operation appears to offer the eventual hope of a near-miraculous cure from their otherwise lethal and crippling illness. If ever the bodies of aborted fetuses were put to good use, it would seem to be here.

It seems frankly impossible to describe these events in morally neutral language. For instance, I am conscious of a heavy irony when I use the phrase 'good use being made' of aborted fetuses. This reflects the unease that is part of a certain kind of moral viewpoint which sees a moral cost in 'making use' of the human body in this way. Indeed, the moral cost of making use of the human body is my main concern in this essay. But notice that seeing a moral cost attaching to something is not the same thing as believing that it should not be done. We can hope that the medical team who are pioneering the technique see the moral costs attaching to the practice of abortion, and perhaps see their technique as somehow contaminated by this moral cost. But if so, it is plain that they see the good which they can do as morally outweighing that cost; confronted by the results of their technique, the dramatic relief of suffering, many might come to the same conclusion. 'Making use' of the human body is justified – if it is justified at all – by the kind of use that is made. On this view, some good uses may really be good enough to provide this justification; some ends may really justify their means.

Now it is important to be clear that there is a distinction between, on the one hand, questions about whether a particular use *is* good enough to offer a justification and, on the other hand, questions about whether 'good uses' *can*, on their own, provide us with moral justification at all. Those who disagree on the first kind of question about a particular case (say, whether the use of fetal brain tissue offers a sufficient good) might nonetheless agree on the second kind of question (concerning whether it is enough to look at the good which is done). They might disagree for instance about whether the long-term prospects for recovery from Parkinsonism are substantial at this early stage in the development of the new technique; they might disagree about the reliability or the generalizability of the results, and so forth. They might even disagree about the significance of the improvement, long-term, even in those sufferers who have been helped. But in all this, they might still agree that it is the *amount of good which is done* which will be decisive as regards whether the new technique can be morally justified.

By contrast, one could hold a quite different view, namely that the amount of good which is done is not the whole story about what is morally right or wrong. Those who think that abortion is *without exception* morally wrong will very likely regard any good actions which follow from an abortion as nonetheless morally unacceptable. Again, those who hold a reverential attitude to the human body, or who have a certain conception of how the dead should be treated with dignity, may regard the harvesting of body parts and tissues as a violation. Equally, those who think that people should not be made use of without their consent may regard the destructive use of the unborn child as thereby morally unacceptable, and so on. My concern is not specifically with whether any of these views is the right one to hold, but rather with being clear about the difference between those who think that the good results hold the key to moral justification, and those who do not. (Of course the way in which we characterize this difference will bear upon the way we would subsequently choose among the moral views which are in front of us.)

I think we could put the contrast like this: those who think that moral justification is provided by good results are in effect saying that we derive our sense of what is morally right from our sense of what is useful, practical, or beneficial; whereas those who hold views such as the ones I have suggested above are, *in all likelihood*, denying this. It is important to qualify the contrast in this way, for the following reason: those who hold what we might call the dissenting views here could do so because they think that certain practices are bound to produce more harm than good in the long run. In other words they might be subscribing to the same general idea that what counts is how much good we do, whilst at the same time fearing the consequences of failing to revere the human body, or of regarding it as a harvestable commodity, and so on. For instance, one sometimes hears the argument that it is wrong to treat the aborted fetus as a source for harvestable tissues because this cheapens our attitude to human life, which will have all sorts of unforeseen harmful consequences; Hitler's Germany is frequently cited as a warning indicator. But it is crucial to see that dissent need not be like this at all: one can straightforwardly deny that we can ever derive what is morally right from what is practically useful or beneficial. As we shall see, such a denial rests on showing that, when we identify which are the good outcomes and which are the harmful ones, we are *already relying on* some conception of what is morally right: hence we could never base an *explanation* of moral judgement on a description or a list of such choices alone.

We do not, however, face a straight choice between the view that good outcomes are the sole determinant of morality, and the view that morality has no place whatsoever for mere practical benefits. If we dislike the idea that the end always justifies the means, we may still think that it can make an important contribution to our moral assessment of them. Experimental treatments, for instance, require close moral scrutiny – but we may nonetheless think that it is morally right to proceed with such an experiment when the patient has no realistic hope of benefitting from the standard alternatives, and has understood this and consented to the experiment. But notice that this always presupposes that the patient has *some* hope of benefitting from the experiment itself: if no good whatsoever could be done by it, the moral justification would fall. Yet by the same token, if the experiment did indeed carry the hope of benefitting the patient, but the risks it involved were concealed from him, the experiment could be morally condemned for reasons that were unconnected with the physical, clinical outcome. Good practical results therefore play some part in moral judgement, but they are not the whole story. Our attitude to the techniques of fetal brain cell transplantation will therefore depend very largely on the relative weight we give to what may be recognized as good outcomes on the one hand, and to other kinds of moral consideration on the other.

It seems, then, that utilitarian perspectives can make only a limited contribution to our understanding of the ownership of human bodies. Put simply, I think that utilitarians must remain silent on the theoretical questions of who, if anyone, *owns* a human body, except insofar as they could claim that ownership questions could be determined by calculations of benefit and harm.[4] On the other hand, utilitarians appear to have much to say on the implementation of various public policies which envisage the *use* of human bodies, their functional capacities, their constituent parts, and their physical products. Some of these uses certainly raise questions of ownership (consider for instance the famous dispute over the ownership of a cell-line which began with diseased tissue being removed for biopsy purposes and ended with a lucrative pharmaceutical development). Since utilitarians are concerned to maximize the practical benefits or good outcomes from actions and policies, they will be anxious to discuss, to analyze and – in some instances – to promote many practices in which human bodies and human products are used. Here are some which spring readily to mind: obtaining, storage, and retrieval of human gametes and embryos; the adoption and fostering of children; physical labor in either paid employment or slavery; sexual prostitution; blood and/or bone marrow donation and sale; organ transplantation; military service and conscription; the use of face and figure in commercial photography, films and advertising; use of the body in medical research and teaching.

Since the aim of these practices is, typically, to produce benefits for *someone or other*, any moral justification for them will most naturally appear to derive from those beneficial outcomes themselves. Consequentialist and especially utilitarian views, which try to derive the morally right from the practically good, might then seem ready-made to offer an account of the moral justification of such practices. We shall turn now to considering two of these practices in a utilitarian light, and to reflecting on whether moral assessment of them involves anything more than counting the goods and the harms involved.

III. TWO PRACTICES CONSIDERED

The Supply of Blood

The demand for blood and blood products is always liable to outstrip supply; the need for blood and blood products is not merely persistent but urgent, given the acute clinical contexts in which it is familiarly used. Accidents and illness can strike anyone, and it is therefore in everyone's interests that an adequate supply be generally available.

A utilitarian would promote those policies for procuring blood which most effectively matched need to potential demand, those policies in effect which maximized the supply. Many developed countries presently rely on a generally voluntary system of *unpaid* blood donation. Alternative systems which rewarded some or all blood donation would preserve at least the appearance of voluntariness; more radical alternatives would abandon voluntariness in favor of compulsory blood procurement. It is possible to imagine a mixture of policies, where voluntary free donations of blood are supplemented by rewarded blood procurement; it is more difficult to imagine a genuine mixture of voluntary and compulsory blood procurement in the same community.

Which of these policies or mixtures of policies would a utilitarian advocate? There is probably no single clear-cut answer to this, since the utilitarian would be guided by whichever system was, in practice, most likely to provide a maximal supply of blood *consistent with other goods and benefits*; and we cannot say *a priori* which system will do this. It is important to remember that the utilitarian need not and probably cannot concern himself with one goal in isolation, even so vital a goal as that of maximizing the supply of blood. Since the utilitarian is committed to those actions and policies which produce the greatest *aggregated* net human benefit, he is bound to take into account the effects which a blood procurement policy might have on other areas of human activity (which after all partly underwrite the importance of preserving and saving lives, to which blood procurement is harnessed).[5] Thus the various effects and consequences of the different policies for blood procurement have to be anticipated and weighed. For this reason, the utilitarian is not committed – as he might otherwise have been – to the simple expedient of forcibly imprisoning and maintaining, as living blood banks, just sufficient human individuals (for instance, violent offenders or the mentally incompetent) as will provide the required amounts of blood which, statistically, the rest of society is likely to need. The revulsion which most would feel at such policies is a consequence which, if not exactly like any other, can and must be weighed by the utilitarian. Short of such a policy, then, what might the utilitarian consider?

The foremost champion of all-voluntary donor systems has been Richard Titmuss, who identifies a number of harms arising from the paid procurement of blood, prominent among them being an expected reduction in the amount of blood obtained voluntarily and a lowering of the quality and safety of the blood so obtained ([9], especially Chapter 8). Critics have disputed these claims and have argued that the burden of quality assurance rests upon the way in which blood donors are recruited and the blood screened and conserved, rather than on the motives of the donors: all markets, after all, presuppose *some* commitment to quality assurance, if not actually to trust [1]. The utilitarian can consider these arguments in terms of their consequences alone, of course. Among the less tangible of these is the question of whether a mixed or rewarded system of blood procurement will depress the generous impulses of those who would, in an all-donor system, give their blood freely and for no reward. For instance, people who would otherwise be willing to give their blood freely may be less willing to do so if they fear that it is likely to become or to supplement what is in effect a commercial commodity; or alternatively, if they feel an injustice in the fact that some but not all giving of blood is rewarded; or for still other reasons [7,9]. Few would contest the view that, when other things are equal, the world is a better place if people give expression to their generous impulses, and a worse place if they restrain them – for whatever reason. Precisely to quantify generosity seems to be a characteristically hopeless task, typical of the kind of calculation to which the utilitarian is strictly committed. Nonetheless the utilitarian can *claim* that the effect of procurement policies on the extent to which generous-minded people give practical expression to their generosity, is an outcome which could in principle be measured, and as such can be accommodated and described in a utilitarian perspective.

So far I am assuming that those who receive payment for their blood are volunteers in the ordinary sense of the word, namely that they freely choose their course of action and are not coerced into it. However this assumption is implicitly disputed in a further objection to rewarded blood donation, which regards even a legally-regulated system of payment for blood as at least potentially coercive of the poor.[6] Even where regulation took this into account, it might be objected that introducing reward into a previously altruistic context constrains the choices of *all* blood donors, rewarded or otherwise. Again, the utilitarian might *claim* to be able to accommodate such objections under a general classification of measurable outcomes.

It seems that we could summarize the arguments concerning the regularized sale of blood like this:

For	*Against*
• we maximize the freedom or entitlement of the moral agent; as they say, it's 'my' body. (But on this last matter, I contend, utilitarians must remain largely silent, though they can urge maximization of freedom as a good in itself or as maximizing the expression or enactment of preferences);	• altruism is discouraged (for instance it is incommensurate with a rewarded system). As such we hinder the freedom of the moral agent;
• the total of blood/plasma/products is increased;	• the supply suffers reduction in quality;
• we reduce exploitative pressure on imports from unregulated commercial suppliers of blood-for-cash (especially in the third world). This of course is true only if the previous point about increased total stocks is true.	• we risk exploiting new vulnerable groups in the society which adopts rewarded procurement
	• we undermine the idea of civic duty, which may but need not find expression in 'sheer' altruism.

Weighing up these various arguments could be presented as a characteristically consequentialist exercise, in that it is consequences which are weighed. Nonetheless, we may still be left with the need to *explain* why some of the consequences are to be given weight. The ideas of altruism and of civic duty on the one hand and of exploitation on the other seem resistant to being themselves cashed out in terms only of further consequences. We could be concerned about the sheer idea of sale *per se* in connection with human tissues and organs. Furthermore, we might distinguish fears that a mixture of gift and rewarded donation would be merely *impractical* from fears that it would be wrong in principle, by *tainting* the gift. These latter fears cannot be further reduced to component consequences. They are sometimes expanded by considering the *quality of the act* of giving, whereby intentions become relevant (and even predominant) in moral assessment. Some moral arguments against sexual prostitution are of this kind. Such considerations are, I think, typically beyond the reach of consequentialist analysis. In sum, there are limits to what can be refuted by an appeal to consequences alone, and these limits are certainly reached in considering the effect of introducing monetary reward into all-volunteer systems of blood donation.

The radical alternative to any volunteer system, paid or unpaid, is one of compulsory procurement (clearly the word 'donation' is entirely out of place here.)[7] Since blood is a good which any might need, yet which nearly all of us can to some extent yield up and replenish without harm to ourselves, it could arguably be considered in much the same light as income for the purposes of taxation. Robert Nozick apart, even conservatives are reticent about disputing the legitimacy of taxation for social provision (that is, taxation beyond the amount necessary for the so-called 'minimal state' provision of national defence and civil order). The social or welfare proportion of taxation can be regarded as a compulsory procurement of money to supplement or even largely replace charitable giving. Whether the amount of money people will give to charity, and hence the sum total of their altruism, is lessened by their paying income tax is unclear. However even if it is, this consideration is, I suppose, defeated by the sheer necessity of maintaining our social obligations towards the neediest members of the community. If this is true of our income, might it not also be true in an exactly parallel sense of our blood?

I don't intend to pursue the details of the case here. The importance of this question concerns whether the utilitarian need pursue it, since at any rate it suggests the characteristically utilitarian priority of the general good as against any particular individual good. While someone *might* claim that this priority worked in the opposite direction, such that a social interest in avoiding the alarm and resentment associated with compulsory blood procurement outweighed specific individual interests in receiving an assured blood supply for transfusion purposes, such a claim would I think fail. It would fail because our interest in not being subjected to compulsory blood procurement is marginal compared to our perhaps life-or-death interest in being able to receive blood and blood products when we need them. (Modern utilitarians are familiar with the requirement of giving due weight to the gravest and most urgent needs such as the need for life-saving or life-maintaining medical treatment, and to supporting this weight even in hypothetical or probabilistic assessments, such as the likelihood that we will actually need such treatment.) If the utilitarian is concerned to promote the general as opposed to the specific individual good, then he may have good reason to take the benefits of compulsory blood procurement seriously, and of course good reason to weigh these benefits very carefully against the harms that might be imagined from such a policy.

It is worth noting here as elsewhere that one need not be a utilitarian in order to favor a system of compulsory procurement. One could deny that the utilitarian has a monopoly over (or even any share whatever in) the wisdom of promoting the general good. Highly *un*consequentialist notions of civic virtue and civic duty may convince me that as a moral agent I have responsibilities to act in such a way as to contribute to the supply of available blood, and that these responsibilities ought to be generalized in terms of legal obligation, as part of the expression of good citizenship. What is at stake here is promoting good actions rather than simply good outcomes.

So far in this discussion I have left out entirely the question of who *owns* the blood we need. The reason is simple: there is no obvious answer, and we are no nearer an answer when we decide how *useful* it would be if, for instance, it were deemed that society or the community had the title to my blood (or to regular amounts of it). No doubt once it has been procured, blood *becomes* the property of the procuring agency, either by gift or by sale. Again, the question of the title to it is slightly different from the question of the rights of disposal. I might conceivably make a donation of my blood conditional on its being put only to certain uses (though whether any actual blood procurement agency would accept such conditions is another matter). What is at any rate clear is that any argument, which merely *stipulated* a social title to my blood as the basis of a policy of compulsory procurement, would be mere rhetoric. The grounds of such a policy would need to be a morally compelling case why we *ought* to accept a general obligation to surrender blood as part of our lawful civic responsibilities, as in income tax. (The state does not need to establish its ownership of my income in order to tax me on it.) By the same token, I could not with any more plausibility base a claim to be paid for giving up my blood on the prior claim that I somehow ‘owned’ it, since we have simply no substantial idea what this means. As Donald Evans has noted, if I retrieve, without permission, a blood sample I have unwillingly yielded to the police for forensic purposes, it is I who am guilty of theft [3]! Any claim to payment would rest on other grounds, such as the inconvenience or risks I incur in giving blood.

The point I am concerned to establish then is that the question of 'ownership' of blood is secondary to and consequent upon the question of how we may best secure a supply of blood and blood products for the medical services which any of us is liable to need. The moral arguments turn on the question of need, the importance of voluntariness and of generosity, and the requirements of obtaining a safe and high quality supply. Seen as *outcomes* of any given policy, each of these will feature in the utilitarian's calculations. But the utilitarian's deliberations will concern how we might best acquire the blood which society needs from me, rather than who (if anyone) owns it while it is still in my veins.

The 'Post Mortem' Examination

There is a familiar sense in which the remains of someone who has died are regarded as 'belonging' to the bereaved family. Whether this sense of belonging signifies *property* is however questionable. It certainly marks out the identity of the deceased – the deceased is the deceased of this bereaved family rather than that one. For instance, most people regard it as of immense importance that they be assured that it is the body of their own loved one, and not mistakenly some other body, which they honor at the ceremony of burial or cremation. Again, it certainly signifies responsibility, for instance the legal responsibility (in the UK at any rate) that the bereaved family have to ensure that the deceased is prepared for and finally committed to burial or cremation in keeping with widely understood standards of public decency and hygiene. By contrast, property considerations seem odd or even bizarre, except in the unusual circumstances where the bereaved family is obstructed in the ordinary process of leave-taking by being prevented from receiving the body of the deceased into their care. The most common reason for such an obstruction is the need for a *post mortem* examination.

We may distinguish three contexts in which *post mortem* examinations may be required: (i) obtaining forensic evidence in the interests of criminal justice; (ii) dispelling uncertainty about the cause of death, often in the interests of public health; (iii) obtaining new knowledge in the pursuit of medical research. Of these contexts, the first two will, in the United Kingdom, characteristically result in *compulsory* removal of the body for *post mortem* examination purposes[8], whereas the third cannot ordinarily be enforced by such compulsion. We can therefore ask what are the moral justifications for compulsion in the first two contexts, and what distinguishes those contexts from the third. Could there be arguments for extending compulsory examination to the context of medical research, again in the public interest? The utilitarian can usually give a substantial account of the outcomes to be balanced when questions of public interest are at issue. But how much do notions of *the body as property* underlie the arguments for or against compulsory removal of the body for examination?

Arguments in favor of compulsion in contexts (i) and (ii) will rely on the strength of the claims to obtain the benefits of carrying out a *post mortem* examination: that is, they aim at identifying an overridingly strong public interest in so doing. The public interest is protected by for instance identifying carriers and routes of infection in the case of serious infectious or contagious diseases, or by identifying, convicting and confining the perpetrators of serious crimes. However there is also a plausible and substantial public interest in the outcome of medical research, in terms of the therapeutic advances at which such research aims. Why then is *this* public interest not decisive in the way that the other two are?

To address this the utilitarian would counterpoise arguments *against* compulsory removal of the body. He might begin by characterizing these in terms of the harmful or distressing effects of such compulsory removal on the bereaved family. For example, objections might center on the violation of dignity (including in some sense the dignity of the deceased) or on intrusion into the privacy of the bereaved family, taking for granted the importance of those parts of any grieving processes which should be private. Further objections might concern the interference with the bodily integrity of the deceased in the context of these grieving processes. These effects can be discerned in the reasons which are given by relatives who refuse permission for a *post mortem* examination for the (non mandatory) purposes of research.[9] Such reasons may arise initially from the timing of a request for permission to carry out a post-mortem: within a day of the death is simply too early for the bereaved relatives to get used to the idea that the person they have lost is now a dead body. Beyond this, the very process of leave-taking is important. Post-mortem examination is widely (and correctly) believed to be a grossly mutilating process and can easily represent both a violation or severance or at best gross interruption to the process of leave-taking or even – in the perception of some relatives, anecdotally – as a further violation of or affront to the deceased: 'He's gone through enough already' is a sometimes-encountered reaction.[10]

(While it is possible to preserve some sort of acceptable appearance to the remains after *post mortem*, by suitable reconstruction and disguise, this seems to represent a deception unlike that involved in the preparations of the undertaker in the ordinary course of events. Whatever the undertaker does is expressly to *facilitate* the process of leave-taking, the personal elements of death, so that they can be accomplished in a safe and seemly way. The *post mortem* by contrast subordinates the personal elements in death to the scientific.[11]

The possibility of these distressing or harmful effects can then, of course, be generalized into a public interest argument in the same way as was considered in connection with compulsory blood procurement: general anxiety or resentment, or alternatively widely-perceived threats against the idea of voluntarism are, for the utilitarian, outcomes to be weighed like any other. At the same time, the utilitarian has not a monopoly on these general outcomes. For instance, it is clear that any of these arguments which express public interests could be remodelled to express *duties*, as was noted in connection with blood donation. For instance, there may be a duty on the part of doctors to *uncover knowledge in the public interest*. It may be that this duty is thought to be decisively strong in those contexts where *post mortem* examination is compulsory, but weaker in the non-compulsory contexts. The utilitarian could retort that the difference between the two kinds of duty devolves onto the different strengths of benefits to be obtained: the benefits of containing presently occurring infectious disease or potentially dangerous criminals currently at large might appear more compelling than those of securing hypothetical gains in future therapeutic methods.

In those contexts where the *post mortem* examination is compulsory, we can infer only that the objections rehearsed here have been overridden, and *not* that they do not arise. In other words, the public interests in contexts (i) and (ii) are regarded as more weighty than the contrary private interests, hence justifying the compulsion. Could or should the compulsion be extended to the context of medical research? The utilitarian's arguments in favor concern maximizing the available benefits to society. If this seems successful in (i) and (ii), what is the decisive difference in context (iii)? Given that the principal objections in all three contexts concern dignity, respect, and voluntarism, why are they successful only in the third?

I am concerned not with what the answer is, but rather with where it may properly be *sought*. Specifically, it is my contention that the answer is not *grounded in* notions of property or 'ownership' of the human body, but rather rests on what are regarded as the decisive moral concerns in each context. Whilst one could subsequently *construct* a notion of 'ownership' in the sense of codifying responsibility for securing and disposing of the body of the deceased, this would be in my view a legal or social fiction that was itself grounded in what we had taken, *for other reasons*, to be the morally appropriate course of action. The body belongs, in this fictional sense, to the family or alternatively to the State by virtue of the rights and responsibilities of disposal. That we ordinarily honor the bereaved family's claim to conduct their grieving and leave-taking in dignity and privacy shows what we ordinarily think is morally compelling, and not who, we think, 'owns' what. Our obligations not to interfere with others go wider than a regard for their property.[12]

Once again, then, my claim is that utilitarian perspectives offer us a description of the things of which we should take account in coming to a moral judgement about *what is the right thing to do*, rather than a means of explicating the idea of property.

IV. CONCLUSION

In the contexts we have been considering and, I think, in the context of other medical uses of the human body and its parts and tissues, what is at issue is not who 'owns' the human body, but rather how we decide what we ought morally to do. To suppose that our moral obligations were here determined by questions of ownership would be to beg the question, and to lose or ignore the richness of our moral language. Indeed, 'ownership' and 'property' are moral notions precisely because more inclusive moral notions such as respect for the worth and dignity of other human beings are prior to them. If we want to make medically advantageous use of the human body we ought to decide, for instance, how to choose between the maximum practical benefit to be obtained on the one hand, and other things that we think important, such as some measure of control over what happens to us, due regard for human dignity, and due respect for people's individuality, on the other.

It is characteristic of this kind of choice in the field of public policy that we face inevitable conflicts between the interests of society as a whole, and the interests of some of its individual members. Society needs blood and other tissues, but individuals must provide them. Society needs access to the dead for forensic and research purposes, but individual families must suffer interruptions to their grieving. Whilst utilitarians can often capture the force of social claims and social interests, these claims and interests can be promoted by a different kind of idea entirely, namely the idea of *civic duty* which individuals can see as decisively urging that they act against their own individual interests on some occasions. Resolving some questions in favor of social claims and against individual interests need not therefore reduce to a utilitarian perspective, although as the proponents of controversial medical technology and research evidently realize, it is utilitarianism which seems to give such claims their most potent guise.

However these moral conflicts are approached, it is clear that they must be understood as such, and not as exercises in semantics. Property claims are no substitute for a wider understanding of welfare, virtue, or duty. Indeed, if we needed to demonstrate property rights in order to sustain an appeal to someone's virtue or duty in the interest of someone else's welfare, the world would be an even more sorry place than it is now.

Centre for Philosophy and Health Care
University of Wales Swansea
United Kingdom

NOTES

1 This point is explored by Griffin, particularly in connection with the promotion of specific alleged 'goods' such as the general phenomenon of promise-keeping ([4], pp. 122-125).

2 Hare's distinction between *intuitive* and *critical* moral thinking fits the distinction between individual moral choices and decisions of public policy very well; it is on the wider level that his prescriptivism is seen as utilitarian [5].

3 Brock cites Brandt and Braybrook in this connection, in the course of his lengthy and instructive overview of modern utilitarian writing ([2], especially pp. 223-224). Those keen to explore modern utilitarianism will find a prodigious bibliography appended to that review.

4 Notions of ownership which were arrived at in this way would seem to be simply constructions. The contrast with our familiar ideas of property is stark: in any ordinary sense, my book is not identified as mine *on the grounds of* the moral implications of my holding it or surrendering it. Rather these moral implications derive from whether or not I own it – quite the reverse of the way the idea of body 'ownership' is derived from what it is morally appropriate to do.

[5] Griffin makes the point that at any rate utilitarianism loses its point unless, as a result of morally good actions, *some* people's lives 'go better'. Unfortunately, in my view, he seems to generalise this so as to become a feature of most views of moral right and wrong (p. 118).

[6] This point is made by Manga against trading in human organs [6].

[7] *En route* to actual compulsion is what is termed by Titmuss the 'Responsibility Fee' donor, whereby patients are loaned the blood they need whilst in hospital, on condition that it is 'repaid' either in blood which they or a proxy subsequently provide, or in the cash equivalent ([9], pp. 78-79).

[8] In England and Wales, application is made to the Coroner, who has authority both to grant a *post mortem* examination and indeed on occasion to direct that one be undertaken. Skegg notes that "a coroner has a prior right to possession of the body when it is required for the purpose of coronial enquiries" [8]. On the limited sense of 'possession' as it is used here, see note 12 below.

[9] I rely here anecdotally on the clinical experience of colleagues.

[10] Whether doctors appreciate this is unclear: it is said to be routine in some British teaching hospitals to ask permission to carry out *post mortem* examinations on most people who die in those hospitals; be it for the purposes of teaching or of research, the examination is nonetheless viewed typically as the completion of the case. At the same time doctors may view a *post mortem* on a patient they have known personally on the ward quite differently from the way they view a dissection, possibly on a long-since dead and almost certainly anonymous body retrieved from a formalin tank.

[11] I am grateful to Dr. David Greaves for this point.

[12] Skegg, perhaps the foremost legal authority on the subject, regards the question of being in lawful possession of a body as entirely distinct from any notion of the body as *property*, and observes that in English law a corpse is specifically regarded as *not* being property [8].

BIBLIOGRAPHY

1. Arrow, K.J.: 1972, 'Gifts and Exchanges', *Philosophy and Public Affairs* **1(4)**, 343-362.
2. Brock, D.W.: 1973, 'Recent work in utilitarianism', *American Philosophical Quarterly* **10(4)**, 241-276.
3. Evans, D.: 1992, *The status and use of body parts and products in the United Kingdom* (unpublished manuscript).
4. Griffin, J.: 1992, 'The human good and the ambitions of consequentialism', *Social Philosophy and Policy* **9(2)**, 118-132.
5. Hare, R.M.: 1981, *Moral thinking: Its levels, method and point*. Clarendon Press, Oxford.
6. Manga, P.: 1987, 'A commercial market for organs? Why not', *Bioethics* **1(4)**, 321-338.
7. Singer, P.: 1973, 'Altruism and commerce: a defense of Titmuss against Arrow', *Philosophy and Public Affairs* **2(4)**, 312-320.
8. Skegg, P.D.G.: 1992, 'Medical uses of corpses and the 'No-Property' Rule', *Medicine, Science and the Law* **32(4)**, 311-318.
9. Titmuss, R.: 1970, *The Gift Relationship: From Human Blood to Social Policy*. George Allen & Unwin, London, UK.

NOTES ON CONTRIBUTORS

Bela Blasszauer, D.J., Ph.D., is Senior Lecturer in Medical Ethics at the Medical University of Pecs, Pecs, Hungary.

Wim J.M. Dekkers, M.D., Ph.D., Department of Ethics, Philosophy and History of Medicine, and Center for Ethics, Catholic University of Nijmegen, Nijmegen, The Netherlands.

Martyn Evans, B.A., Ph.D., is Director of the Centre for Philosophy and Health Care, University of Wales Swansea, Wales, United Kingdom.

Anne Fagot-Largeault, M.D., Ph.D, is Professor of Philosophy, Department of Philosophy, University of Paris-I, Paris, France.

Diego Gracia, M.D., is Professor of History of Medicine and Bioethics, Department of Public Health and History of Science, Complutense University, Madrid, Spain.

Henk A.M.J. ten Have, M.D., Ph.D., is Professor of Medical Ethics, Department of Ethics, Philosophy and History of Medicine, and Center for Ethics, Catholic University of Nijmegen, Nijmegen, The Netherlands.

Dr. med. Friedrich Heubel is Privatdozent für Medizinethik at the Klinikum of Philipps University, Marburg, Germany.

Franz-Joseph Illhardt, D.D., Ph.D., is Executive Director of the Ethics Committee in the Center of Ethics and Law in Medicine, University Clinic Freiburg, Germany.

Uffe J. Jensen, Ph.D., is Professor of Philosophy and Head of the Department of Philosophy; he is also Director of the Research Centre 'Health, Humanity and Culture', Aarhus University, Aarhus, Denmark.

Paul Schotsmans, Ph.D., is Professor of Medical Ethics, Center of Biomedical Ethics and Law, School of Medicine, Catholic University of Louvain, Louvain, Belgium.

Zbigniew Szawarski, Ph.D., is Lecturer at the Centre for Philosophy and Health Care, University of Wales Swansea, Wales, United Kingdom.

H.A.M.J. ten Have and J.V.M. Welie (eds.), Ownership of the Human Body, 227–228.

Jos V.M. Welie, M.Med.S, J.D., Ph.D., is Assistant Professor at the Center for Health Policy and Ethics, Creighton University, Omaha, Nebraska, U.S.A.

Kevin Wm. Wildes, S.J., is Associate Director, The Kennedy Institute of Ethics, Department of Philosophy, Georgetown University, Washington, D.C., U.S.A.

Hub Zwart, Ph.D., is Senior Researcher, Center for Ethics (CEKUN), Catholic University of Nijmegen, Nijmegen, The Netherlands.

INDEX

H.A.M.J. ten Have and J.V.M. Welie (eds.), Ownership of the Human Body, 229–235.

Philosophy and Medicine

1. H. Tristram Engelhardt, Jr. and S.F. Spicker (eds.): *Evaluation and Explanation in the Biomedical Sciences.* 1975 ISBN 90-277-0553-4
2. S.F. Spicker and H. Tristram Engelhardt, Jr. (eds.): *Philosophical Dimensions of the Neuro-Medical Sciences.* 1976 ISBN 90-277-0672-7
3. S.F. Spicker and H. Tristram Engelhardt, Jr. (eds.): *Philosophical Medical Ethics.* Its Nature and Significance. 1977 ISBN 90-277-0772-3
4. H. Tristram Engelhardt, Jr. and S.F. Spicker (eds.): *Mental Health.* Philosophical Perspectives. 1978 ISBN 90-277-0828-2
5. B.A. Brody and H. Tristram Engelhardt, Jr. (eds.): *Mental Illness.* Law and Public Policy. 1980 ISBN 90-277-1057-0
6. H. Tristram Engelhardt, Jr., S.F. Spicker and B. Towers (eds.): *Clinical Judgment.* A Critical Appraisal. 1979 ISBN 90-277-0952-1
7. S.F. Spicker (ed.): *Organism, Medicine, and Metaphysics.* Essays in Honor of Hans Jonas on His 75th Birthday. 1978 ISBN 90-277-0823-1
8. E.E. Shelp (ed.): *Justice and Health Care.* 1981 ISBN 90-277-1207-7; Pb 90-277-1251-4
9. S.F. Spicker, J.M. Healey, Jr. and H. Tristram Engelhardt, Jr. (eds.): *The Law-Medicine Relation.* A Philosophical Exploration. 1981 ISBN 90-277-1217-4
10. W.B. Bondeson, H. Tristram Engelhardt, Jr., S.F. Spicker and J.M. White, Jr. (eds.): *New Knowledge in the Biomedical Sciences.* Some Moral Implications of Its Acquisition, Possession, and Use. 1982 ISBN 90-277-1319-7
11. E.E. Shelp (ed.): *Beneficence and Health Care.* 1982 ISBN 90-277-1377-4
12. G.J. Agich (ed.): *Responsibility in Health Care.* 1982 ISBN 90-277-1417-7
13. W.B. Bondeson, H. Tristram Engelhardt, Jr., S.F. Spicker and D.H. Winship: *Abortion and the Status of the Fetus.* 2nd printing, 1984 ISBN 90-277-1493-2
14. E.E. Shelp (ed.): *The Clinical Encounter.* The Moral Fabric of the Patient-Physician Relationship. 1983 ISBN 90-277-1593-9
15. L. Kopelman and J.C. Moskop (eds.): *Ethics and Mental Retardation.* 1984 ISBN 90-277-1630-7
16. L. Nordenfelt and B.I.B. Lindahl (eds.): *Health, Disease, and Causal Explanations in Medicine.* 1984 ISBN 90-277-1660-9
17. E.E. Shelp (ed.): *Virtue and Medicine.* Explorations in the Character of Medicine. 1985 ISBN 90-277-1808-3
18. P. Carrick: *Medical Ethics in Antiquity.* Philosophical Perspectives on Abortion and Euthanasia. 1985 ISBN 90-277-1825-3; Pb 90-277-1915-2
19. J.C. Moskop and L. Kopelman (eds.): *Ethics and Critical Care Medicine.* 1985 ISBN 90-277-1820-2
20. E.E. Shelp (ed.): *Theology and Bioethics.* Exploring the Foundations and Frontiers. 1985 ISBN 90-277-1857-1
21. G.J. Agich and C.E. Begley (eds.): *The Price of Health.* 1986 ISBN 90-277-2285-4
22. E.E. Shelp (ed.): *Sexuality and Medicine.* Vol. I: Conceptual Roots. 1987 ISBN 90-277-2290-0; Pb 90-277-2386-9

Philosophy and Medicine

23. E.E. Shelp (ed.): *Sexuality and Medicine.* Vol. II: Ethical Viewpoints in Transition. 1987 ISBN 1-55608-013-1; Pb 1-55608-016-6
24. R.C. McMillan, H. Tristram Engelhardt, Jr., and S.F. Spicker (eds.): *Euthanasia and the Newborn.* Conflicts Regarding Saving Lives. 1987 ISBN 90-277-2299-4; Pb 1-55608-039-5
25. S.F. Spicker, S.R. Ingman and I.R. Lawson (eds.): *Ethical Dimensions of Geriatric Care.* Value Conflicts for the 21th Century. 1987 ISBN 1-55608-027-1
26. L. Nordenfelt: *On the Nature of Health.* An Action-Theoretic Approach. 2nd, rev. ed. 1995 ISBN 0-7923-3369-1; Pb 0-7923-3470-1
27. S.F. Spicker, W.B. Bondeson and H. Tristram Engelhardt, Jr. (eds.): *The Contraceptive Ethos.* Reproductive Rights and Responsibilities. 1987 ISBN 1-55608-035-2
28. S.F. Spicker, I. Alon, A. de Vries and H. Tristram Engelhardt, Jr. (eds.): *The Use of Human Beings in Research.* With Special Reference to Clinical Trials. 1988 ISBN 1-55608-043-3
29. N.M.P. King, L.R. Churchill and A.W. Cross (eds.): *The Physician as Captain of the Ship.* A Critical Reappraisal. 1988 ISBN 1-55608-044-1
30. H.-M. Sass and R.U. Massey (eds.): *Health Care Systems.* Moral Conflicts in European and American Public Policy. 1988 ISBN 1-55608-045-X
31. R.M. Zaner (ed.): *Death: Beyond Whole-Brain Criteria.* 1988 ISBN 1-55608-053-0
32. B.A. Brody (ed.): *Moral Theory and Moral Judgments in Medical Ethics.* 1988 ISBN 1-55608-060-3
33. L.M. Kopelman and J.C. Moskop (eds.): *Children and Health Care.* Moral and Social Issues. 1989 ISBN 1-55608-078-6
34. E.D. Pellegrino, J.P. Langan and J. Collins Harvey (eds.): *Catholic Perspectives on Medical Morals.* Foundational Issues. 1989 ISBN 1-55608-083-2
35. B.A. Brody (ed.): *Suicide and Euthanasia.* Historical and Contemporary Themes. 1989 ISBN 0-7923-0106-4
36. H.A.M.J. ten Have, G.K. Kimsma and S.F. Spicker (eds.): *The Growth of Medical Knowledge.* 1990 ISBN 0-7923-0736-4
37. I. Löwy (ed.): *The Polish School of Philosophy of Medicine.* From Tytus Chałubiński (1820–1889) to Ludwik Fleck (1896–1961). 1990 ISBN 0-7923-0958-8
38. T.J. Bole III and W.B. Bondeson: *Rights to Health Care.* 1991 ISBN 0-7923-1137-X
39. M.A.G. Cutter and E.E. Shelp (eds.): *Competency.* A Study of Informal Competency Determinations in Primary Care. 1991 ISBN 0-7923-1304-6
40. J.L. Peset and D. Gracia (eds.): *The Ethics of Diagnosis.* 1992 ISBN 0-7923-1544-8

41. K.W. Wildes, S.J., F. Abel, S.J. and J.C. Harvey (eds.): *Birth, Suffering, and Death.* Catholic Perspectives at the Edges of Life. 1992 [CSiB-1] ISBN 0-7923-1547-2; Pb 0-7923-2545-1
42. S.K. Toombs: *The Meaning of Illness.* A Phenomenological Account of the Different Perspectives of Physician and Patient. 1992 ISBN 0-7923-1570-7; Pb 0-7923-2443-9
43. D. Leder (ed.): *The Body in Medical Thought and Practice.* 1992 ISBN 0-7923-1657-6
44. C. Delkeskamp-Hayes and M.A.G. Cutter (eds.): *Science, Technology, and the Art of Medicine.* European-American Dialogues. 1993 ISBN 0-7923-1869-2
45. R. Baker, D. Porter and R. Porter (eds.): *The Codification of Medical Morality.* Historical and Philosophical Studies of the Formalization of Western Medical Morality in the 18th and 19th Centuries, Volume One: Medical Ethics and Etiquette in the 18th Century. 1993 ISBN 0-7923-1921-4
46. K. Bayertz (ed.): *The Concept of Moral Consensus.* The Case of Technological Interventions in Human Reproduction. 1994 ISBN 0-7923-2615-6
47. L. Nordenfelt (ed.): *Concepts and Measurement of Quality of Life in Health Care.* 1994 [ESiP-1] ISBN 0-7923-2824-8
48. R. Baker and M.A. Strosberg (eds.) with the assistance of J. Bynum: *Legislating Medical Ethics.* A Study of the New York State Do-Not-Resuscitate Law. 1995 ISBN 0-7923-2995-3
49. R. Baker (ed.): *The Codification of Medical Morality.* Historical and Philosophical Studies of the Formalization of Western Morality in the 18th and 19th Centuries, Volume Two: Anglo-American Medical Ethics and Medical Jurisprudence in the 19th Century. 1995 ISBN 0-7923-3528-7; Pb 0-7923-3529-5
50. R.A. Carson and C.R. Burns (eds.): *Philosophy of Medicine and Bioethics.* A Twenty-Year Retrospective and Critical Appraisal. 1997 ISBN 0-7923-3545-7
51. K.W. Wildes, S.J. (ed.): *Critical Choices and Critical Care.* Catholic Perspectives on Allocating Resources in Intensive Care Medicine. 1995 [CSiB-2] ISBN 0-7923-3382-9
52. K. Bayertz (ed.): *Sanctity of Life and Human Dignity.* 1996 ISBN 0-7923-3739-5
53. Kevin Wm. Wildes, S.J. (ed.): *Infertility: A Crossroad of Faith, Medicine, and Technology.* 1996 ISBN 0-7923-4061-2
54. Kazumasa Hoshino (ed.): *Japanese and Western Bioethics.* Studies in Moral Diversity. 1996 ISBN 0-7923-4112-0
55. E. Agius and S. Busuttil (eds.): *Germ-Line Intervention and our Responsibilities to Future Generations.* 1998 ISBN 0-7923-4828-1
56. L.B. McCullough: *John Gregory and the Invention of Professional Medical Ethics and the Professional Medical Ethics and the Profession of Medicine.* 1998 ISBN 0-7923-4917-2
57. L.B. McCullough: *John Gregory's Writing on Medical Ethics and Philosophy of Medicine.* 1998 [CiME-1] ISBN 0-7923-5000-6

Philosophy and Medicine

58. H.A.M.J. ten Have and H.-M. Sass (eds.): *Consensus Formation in Healthcare Ethics*. 1998 [ESiP-2] ISBN 0-7923-4944-X
59. H.A.M.J. ten Have and J.V.M. Welie (eds.): *Ownership of the Human Body*. Philosophical Considerations on the Use of the Human Body and its Parts in Healthcare. 1998 [ESiP-3] ISBN 0-7923-5150-9

KLUWER ACADEMIC PUBLISHERS – DORDRECHT / BOSTON / LONDON